SPACE SCIENCE, EXPLORATION AND POLICIES

FUTURE OF U. S. HUMAN SPACEFLIGHT: BACKGROUND AND ISSUES

SPACE SCIENCE, EXPLORATION AND POLICIES

Additional books in this series can be found on Nova's website under the Series tab.

Additional E-books in this series can be found on Nova's website under the E-book tab.

SPACE SCIENCE, EXPLORATION AND POLICIES

FUTURE OF U. S. HUMAN SPACEFLIGHT: BACKGROUND AND ISSUES

DEREK A. WARREN
AND
BRIDGET D. CONWAY
EDITORS

Nova Science Publishers, Inc.
New York

For permission to use material from this book please contact us:
Telephone 631-231-7269; Fax 631-231-8175
Web Site: http://www.novapublishers.com

LIBRARY OF CONGRESS CATALOGING-IN-PUBLICATION DATA
Future of U.S. human spaceflight : background and issues / editors, Derek A. Warren and Bridget D. Conway.
p. cm.
Includes bibliographical references and index.
ISBN 978-1-61668-774-8 (hardcover)
1. Astronautics--Government policy--United States. 2. Manned space flight--Government policy--United States. I. Warren, Derek A. II. Conway, Bridget D. III. Title: Future of United States human space flight.
TL789.8.U5F793 2009
629.4500973--dc22
2010012196

Published by Nova Science Publishers, Inc. ✝ New York

CONTENTS

PREFACE

Space operations are among the most demanding and unforgiving pursuits ever undertaken by humans and will become all the more difficult when means do not match aspirations. Will humans leave the close proximity of low-Earth orbit, where astronauts have circled since 1972, and explore the solar system, charting a path for the eventual expansion of human civilization into space? If so, how will they ensure that the exploration delivers the greatest benefit to the nation? Can people explore with reasonable assurances of human safety? Can the nation marshal the resources to embark on the mission? This book explores the nations important decisions on the future of human spaceflight. This book consists of public documents which have been located, gathered, combined, reformatted, and enhanced with a subject index, selectively edited and bound to provide easy access.

Chapter 1 - The U.S. human spaceflight program appears to be on an unsustainable trajectory. It is perpetuating the perilous practice of pursuing goals that do not match allocated resources. Space operations are among the most demanding and unforgiving pursuits ever undertaken by humans. It really is rocket science. Space operations become all the more difficult when means do not match aspirations. Such is the case today.

The nation is facing important decisions on the future of human spaceflight. Will humans leave the close proximity of low-Earth orbit, where astronauts have circled since 1972, and explore the solar system, charting a path for the eventual expansion of human civilization into space? If so, how will they ensure that the exploration delivers the greatest benefit to the nation? Can people explore with reasonable assurances of human safety? Can the nation marshal the resources to embark on the mission?

Whatever space program is ultimately selected, it must be matched with the resources needed for its execution. How can we marshal the necessary resources? There are actually more options available today than in 1961, when President Kennedy challenged the nation to "commit itself to the goal, before this decade is out, of landing a man on the Moon and returning him safely to the Earth."

Chapter 2 - The "space age" began on October 4, 1957, when the Soviet Union (USSR) launched Sputnik, the world's first artificial satellite. Some U.S. policymakers, concerned about the USSR's ability to launch a satellite, thought Sputnik might be an indication that the United States was trailing behind the USSR in science and technology. The Cold War also led some U.S. policymakers to perceive the Sputnik launch as a possible precursor to nuclear attack. In response to this "Sputnik moment," the U.S. government undertook several policy actions, including the establishment of the National Aeronautics and Space Administration

(NASA) and the Defense Advanced Research Projects Agency (DARPA), enhancement of research funding, and reformation of science, technology, engineering and mathematics (STEM) education policy.

Following the "Sputnik moment," a set of fundamental factors gave "importance, urgency, and inevitability to the advancement of space technology," according to an Eisenhower presidential committee. These four factors include the compelling need to explore and discover; national defense; prestige and confidence in the U.S. scientific, technological, industrial, and military systems; and scientific observation and experimentation to add to our knowledge and understanding of the Earth, solar system, and universe. They are still part of current policy discussions and influence the nation's civilian space policy priorities—both in terms of what actions NASA is authorized to undertake and the appropriations each activity within NASA receives.

Further, the United States faces a far different world today. No Sputnik moment, Cold War, or space race exists to help policymakers clarify the goals of the nation's civilian space program. The Hubble telescope, *Challenger* and *Columbia* space shuttle disasters, and Mars exploration rovers frame the experience of current generations, in contrast to the Sputnik launch and the U.S. Moon landings. As a result, some experts have called for new 21st century space policy objectives and priorities to replace those developed 50 years ago.

On October 15, 2008, the NASA Authorization Act of 2008 (P.L. 110-422) was signed into law. This act authorized appropriations for FY2009, and prohibited NASA from taking any steps prior to April 30, 2009, that would preclude the President and Congress from being able to continue to fly the Space Shuttle past 2010. During the 111th Congress, policymakers may discuss another authorization bill for future years, and identify priorities for civil space exploration. Little is known about the Obama Administration's space policy priorities, and the degree to which they agree with Bush Administration policies.

The "space age" began on October 4, 1957, when the Soviet Union (USSR) launched Sputnik, the world's first artificial satellite. Some U.S. policymakers, concerned about the USSR's ability to launch a satellite, thought Sputnik might be an indication that the United States was trailing behind the USSR in science and technology. The Cold War also led some U.S. policymakers to perceive the Sputnik launch as a possible precursor to nuclear attack. In response to this "Sputnik moment," the U.S. government undertook several policy actions, including the establishment of the National Aeronautics and Space Administration (NASA) and the Defense Advanced Research Projects Agency (DARPA), enhancement of research funding, and reformation of science, technology, engineering and mathematics (STEM) education policy.

Following the "Sputnik moment," a set of fundamental factors gave "importance, urgency, and inevitability to the advancement of space technology," according to an Eisenhower presidential committee. These four factors include the compelling need to explore and discover; national defense; prestige and confidence in the U.S. scientific, technological, industrial, and military systems; and scientific observation and experimentation to add to our knowledge and understanding of the Earth, solar system, and universe. They are still part of current policy discussions and influence the nation's civilian space policy priorities—both in terms of what actions NASA is authorized to undertake and the appropriations each activity within NASA receives.

Further, the United States faces a far different world today. No Sputnik moment, Cold War, or space race exists to help policymakers clarify the goals of the nation's civilian space program. The Hubble telescope, *Challenger* and *Columbia* space shuttle disasters, and Mars exploration rovers frame the experience of current generations, in contrast to the Sputnik launch and the U.S. Moon landings. As a result, some experts have called for new 21st century space policy objectives and priorities to replace those developed 50 years ago.

On October 15, 2008, the NASA Authorization Act of 2008 (P.L. 110-422) was signed into law. This act authorized appropriations for FY2009, and prohibited NASA from taking any steps prior to April 30, 2009, that would preclude the President and Congress from being able to continue to fly the Space Shuttle past 2010. During the 111th Congress, policymakers may discuss another authorization bill for future years, and identify priorities for civil space exploration. Little is known about the Obama Administration's space policy priorities, and the degree to which they agree with Bush Administration policies.

Chapter 3 - The International Space Station (ISS) program began in 1993, with Russia joining the United States, Europe, Japan, and Canada. Crews have occupied ISS on a 4-6 month rotating basis since November 2000.

The U.S. Space Shuttle, which first flew in April 1981, has been the major vehicle taking crews and cargo back and forth to ISS, but the shuttle system has encountered difficulties since the *Columbia* disaster in 2003. Russian Soyuz spacecraft are also used to take crews to and from ISS, and Russian Progress spacecraft deliver cargo, but cannot return anything to Earth, since they are not designed to survive reentry into the Earth's atmosphere. A Soyuz is always attached to the station as a lifeboat in case of an emergency.

President Bush, prompted in part by the *Columbia* tragedy, made a major space policy address on January 14, 2004, directing NASA to focus its activities on returning humans to the Moon and someday sending them to Mars. Included in this "Vision for Space Exploration" is a plan to retire the space shuttle in 2010. The President said the United States would fulfill its commitments to its space station partners, and the shuttle *Discovery* made the first post-*Columbia* flight to the ISS in July 2006. Shuttle flights have continued and completion of the space station is scheduled before the shuttle is retired in 2010. Meanwhile NASA has begun development of a new crew launch vehicle, named Ares, and a crew exploration vehicle, named Orion.

NASA programs were funded for FY2008 in Division B of the Consolidated Appropriations Act (P.L. 110-161). The Space Operations program, which includes the space shuttle and the ISS, was funded at $6.734 billion. For FY2009 NASA requested $5.775 billion for these programs, but in the process revised its budgeting to move its overhead costs to a new account called Cross- Agency Support. Under the new system, the FY2008 Space Operations program would have received $5.526 billion, about $250 million less than the FY2009 request. NASA is currently operating under a continuing resolution (Division A of P.L. 110-329), which funded most civilian activities through March 6, 2009. Under the continuing resolution, Space Operations are funded at the $5.526 billion rate appropriated for FY2008.

An FY2009 NASA authorization bill (H.R. 6063) was introduced May 15, 2008. Among the provisions in the one-year authorization bill was a "Sense of the Congress" urging cooperation in the Moon/Mars activities with other nations pursuing human space flight. It also requires that NASA "terminate or suspend any activity of the Agency that, if continued between the date of enactment of this Act and April 30, 2009, would preclude the continued safe and effective flight of the Space Shuttle after fiscal year 2010 if the President inaugurated on January 20, 2009, were to make a determination to delay the Space Shuttle's scheduled retirement." Congress passed the bill September 27, and it was signed by the President October 15 (P.L. 110-422).

Chapters 4, 5, 6 and 7 feature testimony before the U. S. House of Representatives.

In: Future of U.S. Human Spaceflight: Background and Issues ISBN: 978-1-61668-774-8
Ed: Derek A. Warren and Bridget D. Conway

Chapter 1

SEEKING A HUMAN SPACEFLIGHT PROGRAM WORTHY OF A GREAT NATION*

Review of United States Human Spaceflight Plans Committee

EXECUTIVE SUMMARY

The U.S. human spaceflight program appears to be on an unsustainable trajectory. It is perpetuating the perilous practice of pursuing goals that do not match allocated resources. Space operations are among the most demanding and unforgiving pursuits ever undertaken by humans. It really is rocket science. Space operations become all the more difficult when means do not match aspirations. Such is the case today.

The nation is facing important decisions on the future of human spaceflight. Will we leave the close proximity of low-Earth orbit, where astronauts have circled since 1972, and explore the solar system, charting a path for the eventual expansion of human civilization into space? If so, how will we ensure that our exploration delivers the greatest benefit to the nation? Can we explore with reasonable assurances of human safety? Can the nation marshal the resources to embark on the mission?

Whatever space program is ultimately selected, it must be matched with the resources needed for its execution. How can we marshal the necessary resources? There are actually more options available today than in 1961, when President Kennedy challenged the nation to "commit itself to the goal, before this decade is out, of landing a man on the Moon and returning him safely to the Earth."

First, space exploration has become a global enterprise. Many nations have aspirations in space, and the combined annual budgets of their space programs are comparable to NASA's. If the United States is willing to lead a global program of exploration, sharing both the burden and benefit of space exploration in a meaningful way, significant accomplishments could follow. Actively engaging international partners in a manner adapted to today's multi-polar

* This is an edited, reformatted and augmented version of a U. S. Human Spaceflight Plans Committee.

world could strengthen geopolitical relationships, leverage global financial and technical resources, and enhance the exploration enterprise.

Second, there is now a burgeoning commercial space industry. If we craft a space architecture to provide opportunities to this industry, there is the potential—not without risk—that the costs to the government would be reduced. Finally, we are also more experienced than in 1961, and able to build on that experience as we design an exploration program. If, after designing cleverly, building alliances with partners, and engaging commercial providers, the nation cannot afford to fund the effort to pursue the goals it would like to embrace, it should accept the disappointment of setting lesser goals.

Can we explore with reasonable assurances of human safety? Human space travel has many benefits, but it is an inherently dangerous endeavor. Human safety can never be absolutely assured, but throughout this report, safety is treated as a *sine qua non*. It is not discussed in extensive detail because any concepts falling short in human safety have simply been eliminated from consideration.

How will we explore to deliver the greatest benefit to the nation? Planning for a human spaceflight program should begin with a choice about its goals—rather than a choice of possible destinations. Destinations should derive from goals, and alternative architectures may be weighed against those goals. There is now a strong consensus in the United States that the next step in human spaceflight is to travel beyond low-Earth orbit. This should carry important benefits to society, including: driving technological innovation; developing commercial industries and important national capabilities; and contributing to our expertise in further exploration. Human exploration *can* contribute appropriately to the expansion of scientific knowledge, particularly in areas such as field geology, and it is in the interest of both science and human spaceflight that a credible and well-rationalized strategy of coordination between them be developed. Crucially, human spaceflight objectives should broadly align with key national objectives.

These more tangible benefits exist within a larger context. Exploration provides an opportunity to demonstrate space leadership while deeply engaging international partners; to inspire the next generation of scientists and engineers; and to shape human perceptions of our place in the universe. The Committee concludes that the ultimate goal of human exploration is to chart a path for human expansion into the solar system. This is an ambitious goal, but one worthy of U.S. leadership in concert with a broad range of international partners.

The Committee's task was to review the U.S. plans for human spaceflight and to offer possible alternatives. In doing so, it assessed the programs within the current human spaceflight portfolio; considered capabilities and technologies a future program might require; and considered the roles of commercial industry and our international partners in this enterprise. From these deliberations, the Committee developed five integrated alternatives for the U.S. human spaceflight program, including an executable version of the current program. The considerations and the five alternatives are summarized in the pages that follow.

Key Questions to Guide the Plan for Human Spaceflight

The Committee identified the following questions that, if answered, would form the basis of a plan for U.S. human spaceflight:

1. What should be the future of the Space Shuttle?
2. What should be the future of the International Space Station (ISS)?
3. On what should the next heavy-lift launch vehicle be based?
4. How should crews be carried to low-Earth orbit?
5. What is the most practicable strategy for exploration beyond low-Earth orbit?

The Committee considers the framing and answering of these questions individually and consistently to be at least as important as their combinations in the integrated options for a human spaceflight program, which are discussed below. Some 3,000 alternatives can be derived from the various possible answers to these questions; these were narrowed to the five representative families of integrated options that are offered in this report. In these five families, the Committee examined the interactions of the decisions, particularly with regard to cost and schedule. Other reasonable and consistent combinations of the choices are possible (each with its own cost and schedule implications), and these could also be considered as alternatives.

Current Programs

Before addressing options for the future human exploration program, it is appropriate to discuss the current programs: the Space Shuttle, the International Space Station and Constellation, as well as the looming problem of "the gap"—the time that will elapse between the scheduled completion of the Space Shuttle program and the advent of a new U.S. capability to lift humans into space.

Space Shuttle

What should be the future of the Space Shuttle? The current plan is to retire it at the end of FY 2010, with its final flight scheduled for the last month of that fiscal year. Although the current administration has relaxed the requirement to complete the last mission before the end of FY 2010, there are no funds in the FY 2011 budget for continuing Shuttle operations.

In considering the future of the Shuttle, the Committee assessed the realism of the current schedule; examined issues related to the Shuttle workforce, reliability and cost; and weighed the risks and possible benefits of a Shuttle extension. The Committee noted that the projected flight rate is nearly twice that of the actual flight rate since return to flight in 2005 after the Columbia accident two years earlier. Recognizing that undue schedule and budget pressure can subtly impose a negative influence on safety, the Committee finds that a more realistic schedule is prudent. With the remaining flights likely to stretch into the second quarter of FY 2011, the Committee considers it important to budget for Shuttle operations through that time.

Although a thorough analysis of Shuttle safety was not part of its charter, the Committee did examine the Shuttle's safety record and reliability, as well as the results of other reviews of these topics. New human-rated launch vehicles will likely be more reliable once they reach maturity, but in the meantime, the Shuttle is in the enviable position of being through its "infant mortality" phase. Its flight experience and demonstrated reliability should not be discounted.

Once the Shuttle is retired, there will be a gap in the capability of the United States itself to launch humans into space. That gap will extend until the next U.S. human-rated launch system becomes available. The Committee estimates that, under the current plan, this gap will be at least seven years. There has not been this long a gap in U.S. human launch capability since the U.S. human space program began.

Most of the integrated options presented below would retire the Shuttle after a prudent fly-out of the current manifest, indicating that the Committee found the interim reliance on international crew services acceptable. However, one option does provide for an extension of the Shuttle at a minimum safe flight rate to preserve U.S. capability to launch astronauts into space. If that option is selected, there should be a thorough review of Shuttle recertification and overall Shuttle reliability to ensure that the risk associated with that extension would be acceptable. The results of the recertification should be reviewed by an independent committee, with the purpose of ensuring that NASA has met the intent behind the relevant recommendation of the Columbia Accident Investigation Board.

International Space Station

In considering the future of the International Space Station, the Committee asked two basic questions: What is the outlook between now and 2015? Should the ISS be extended beyond 2015?

The Committee is concerned that the ISS, and particularly its utilization, may be at risk after Shuttle retirement. The ISS was designed, assembled and operated with the capabilities of the Space Shuttle in mind. The present approach to its utilization is based on Shuttle-era experience. After Shuttle retirement, the ISS will rely on a combination of new international vehicles and as-yet-unproven U.S. commercial vehicles for cargo transport. Because the planned commercial resupply capability will be crucial to both ISS operations and utilization, it may be prudent to strengthen the incentives to the commercial providers to meet the schedule milestones.

Now that the ISS is nearly completed and is staffed by a full crew of six, its future success will depend on how well it is used. Up to now, the focus has been on assembling the ISS, and this has come at the expense of exploiting its capabilities. Utilization should have first priority in the years ahead.

The Committee finds that the return on investment from the ISS to both the United States and the international partners would be significantly enhanced by an extension of its life to 2020. It seems unwise to de-orbit the Station after 25 years of planning and assembly and only five years of operational life. A decision *not* to extend its operation would significantly impair the U.S. ability to develop and lead future international spaceflight partnerships. Further, the return on investment from the ISS would be significantly increased if it were funded at a level allowing it to achieve its full potential: as the nation's newest National Laboratory, as an enhanced testbed for technologies and operational techniques that support exploration, and as a management framework that can support expanded international collaboration.

The strong and tested working relationship among international partners is perhaps the most important outcome of the ISS program. The partnership expresses a "first among equals" U.S. leadership style adapted to today's multi-polar world. That leadership could extend to exploration, as the ISS partners could engage at an early stage if aspects of exploration beyond low-Earth orbit were included in the goals of the partnership agreement. (See Figure i.)

The Constellation Program

The Constellation Program includes the Ares I launch vehicle, capable of launching astronauts to low-Earth orbit; the Ares V heavy-lift launch vehicle, to send astronauts and equipment to the Moon; the Orion capsule, to carry astronauts to low-Earth orbit and beyond; and the Altair lunar lander and lunar surface systems astronauts will need to explore the lunar surface. As the Committee assessed the current status and possible future of the Constellation Program, it reviewed the technical, budgetary, and schedule challenges that the program faces today.

Given the funding upon which it was based, the Constellation Program chose a reasonable architecture for human exploration. However, even when it was announced, its budget depended on funds becoming available from the retirement of the Space Shuttle in 2010 and the decommissioning of ISS in early 2016. Since then, as a result of technical and budgetary issues, the development schedules of Ares I and Orion have slipped, and work on Ares V and Altair has been delayed.

Most major vehicle-development programs face technical challenges as a normal part of the process, and Constellation is no exception. While significant, these are engineering problems that the Committee expects can be solved. But these solutions may add to the program's cost and delay its schedule.

The original 2005 schedule showed Ares I and Orion available to support the ISS in 2012, two years after scheduled Shuttle retirement. The current schedule now shows that date as 2015. An independent assessment of the technical, budgetary and schedule risk to the Constellation Program performed for the Committee indicates that an additional delay of at least two years is likely. This means that Ares I and Orion will not reach the ISS before the Station's currently planned termination, and the length of the gap in U.S. ability to launch astronauts into space will be at least seven years.

The Committee also examined the design and development of Orion. Many concepts are possible for crew-exploration vehicles, and NASA clearly needs a new spacecraft for travel beyond low-Earth orbit. The Committee found no compelling evidence that the current design will not be acceptable for its wide variety of tasks in the exploration program. However, the Committee is concerned about Orion's recurring costs. The capsule is considerably larger and more massive than previous capsules (e.g., the Apollo capsule), and there is some indication that a smaller and lighter four- person Orion could reduce operational costs. However, a redesign of this magnitude would likely result in more than a year of additional development time and a significant increase in development cost, so such a redesign should be considered carefully before being implemented.

Capability for Launch to Low-Earth Orbit and Exploration beyond

Heavy-lift launch to low-Earth orbit and beyond

No one knows the mass or dimensions of the largest hardware that will be required for future exploration missions, but it will likely be significantly larger than 25 metric tons (mt) in launch mass to low-Earth orbit, which is the capability of current launchers. As the size of the launcher increases, the result is fewer launches and less operational complexity in terms of assembly and/or refueling in space. In short, the net availability of launch capability increases.

Combined with considerations of launch availability and on-orbit operations, the Committee finds that exploration would benefit from the availability of a heavy-lift vehicle. In addition, heavy-lift would enable the launching of large scientific observatories and more capable deep-space missions. It may also provide benefit in national security applications. The question this raises is: On what system should the next heavy-lift launch vehicle be based?

Potential approaches to developing heavy-lift vehicles are based on NASA heritage (Shuttle and Apollo) and (EELV) Evolved Expendable Launch Vehicle heritage. (See Figure ii.) Each has distinct advantages and disadvantages. In the Ares-V-plus-Ares-I system planned by the Constellation Program, the Ares I launches the Orion and docks in low- Earth orbit with the Altair lander launched on the Ares V. This configuration has the advantage of projected very high ascent crew safety, but it delays the development of the Ares V heavy-lift vehicle until after the Ares I is developed.

In a different, related architecture, the Orion and Altair are launched on two separate "Lite" versions of the Ares V, providing for more robust mission mass and volume margins. Building a single NASA vehicle could reduce carrying and operations costs and accelerate heavy-lift development. Of these two Ares system alternatives, the Committee finds the Ares V Lite used in the dual mode for lunar missions to be the preferred reference case.

The Shuttle-derived family consists of in-line and side-mount vehicles substantially derived from the Shuttle, thereby providing greater workforce continuity. The development cost of the more Shuttle-derived system would be lower, but it would be less capable than the Ares V family and have higher recurring costs. The lower lift capability could eventually be offset by developing on-orbit refueling.

The EELV-heritage systems have the least lift capacity, requiring almost twice as many launches as the Ares family to attain equal performance. If on-orbit refueling were developed and used, the number of launches could be reduced, but operational complexity would increase. However, the EELV approach would also represent a new way of doing business for NASA, which would have the benefit of potentially lowering development and operational costs. This would come at the expense of ending a substantial portion of the internal NASA capability to develop and operate launchers. It would also require that NASA and the Department of Defense jointly develop the new system.

Family			Launch Mass to LEO
NASA Heritage	Ares Family	Ares V + Ares I	160 mt + 25 mt
		Ares V Lite	140 mt
	Shuttle Derived Family		100 -110 mt
EELV Heritage Family			75 mt

Source: Review of U.S. Human Spaceflight Plans Committee

Figure i. Characteristics of heavy-lift launch vehicles, indicating the EELV and NASA heritage families

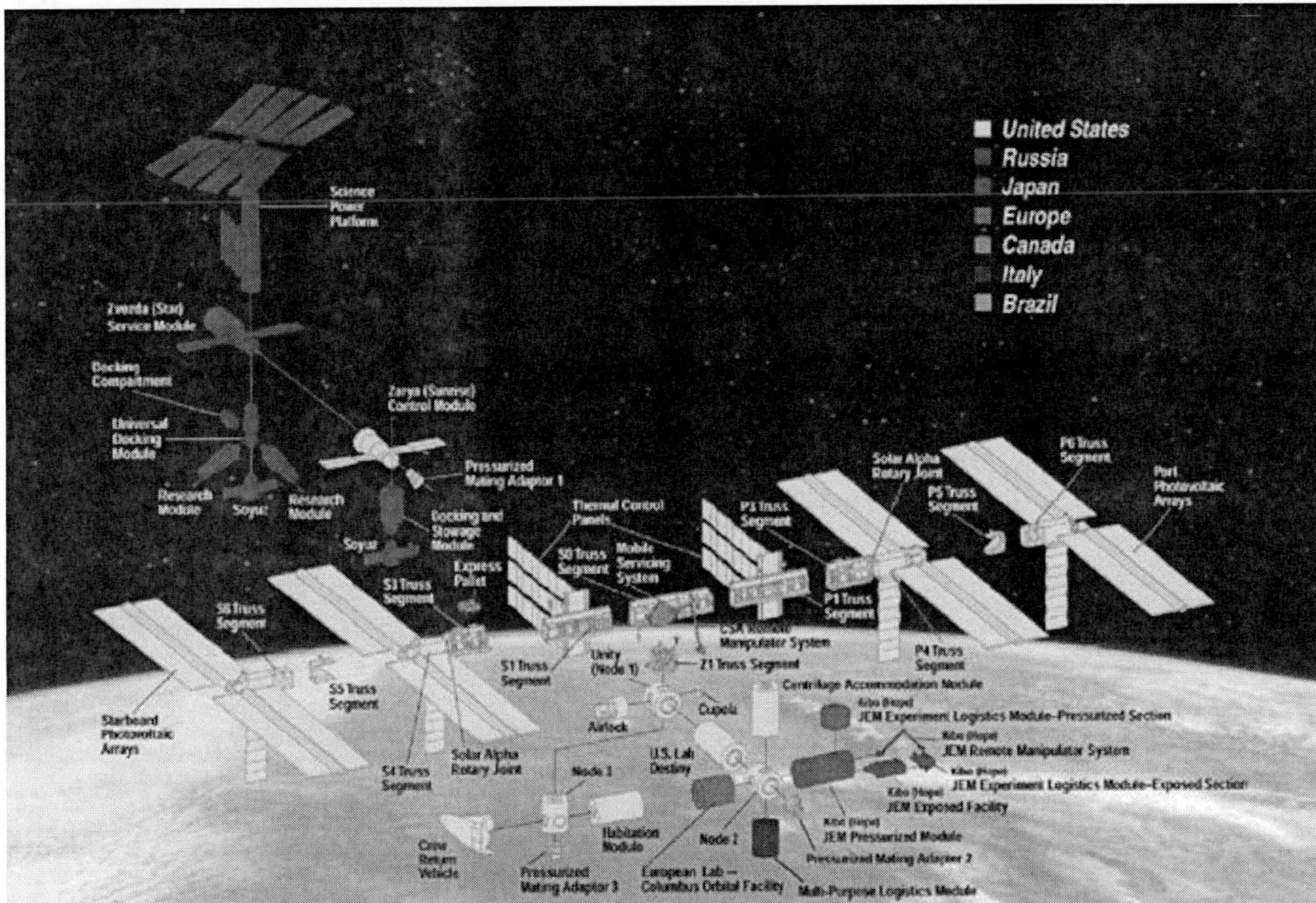

Source: NASA

Figure ii. Diagram of the International Space Station showing elements provided by each of the international partners.

All of the options would benefit from the development of in-space refueling, and the smaller rockets would benefit most of all. A potential government-guaranteed market to provide fuel in low-Earth orbit would create a strong stimulus to the commercial launch industry.

The Committee cautions against the tradition of designing for ultimate performance at the expense of reliability, operational efficiency, and life-cycle cost.

Crew access to low-Earth Orbit

How should U.S. astronauts be transported to low-Earth orbit? There are two basic approaches: a government- operated system and a commercial transport service. The current Constellation Program plan is to use the government-operated Ares I launch vehicle and the Orion crew capsule. However, the Committee found that, because of technical and budget issues, the Ares I schedule no longer supports ISS needs.

Ares I was designed to a high safety standard to provide astronauts with access to low-Earth orbit at lower risk and a considerably higher level of safety than is available today. To achieve this, it uses a high-reliability rocket and a crew capsule with a launch-escape system. But other combinations of high-reliability rockets and capsules with escape systems could also provide that safety. The Committee was unconvinced that enough is known about any of the potential high-reliability launcher-plus-capsule systems to distinguish their levels of safety in a meaningful way.

The United States needs a means of launching astronauts to low-Earth orbit, but it does not necessarily have to be provided by the government. As we move from the complex, reusable Shuttle back to a simpler, smaller capsule, it is appropriate to consider turning this transport service over to the commercial sector. This approach is not without technical and programmatic risks, but it creates the possibility of lower operating costs for the system and potentially accelerates the availability of U.S. access to low-Earth orbit by about a year, to 2016. If this option is chosen, the Committee suggests establishing a new competition for this service, in which both large and small companies could participate.

Lowering the cost of space exploration

The cost of exploration is dominated by the costs of launch to low-Earth orbit and of in-space systems. It seems improbable that significant reductions in launch costs will be realized in the short term until launch rates increase substantially—perhaps through expanded commercial activity in space. How can the nation stimulate such activity? In the 1920s, the federal government awarded a series of guaranteed contracts for carrying airmail, stimulating the growth of the airline industry. The Committee concludes that an exploration architecture employing a similar policy of guaranteed contracts has the potential to stimulate a vigorous and competitive commercial space industry. Such commercial ventures could include the supply of cargo to the ISS (planning for which is already under way by NASA and industry – see Figure iii), transport of crew to orbit and transport of fuel to orbit. Establishing these commercial opportunities could increase launch volume and potentially lower costs to NASA and all other launch services customers. This would have the additional benefit of focusing NASA on a more challenging role, permitting it to concentrate its efforts where its inherent capability resides: in developing cutting-edge technologies and concepts, defining programs, and overseeing the development and operation of exploration systems.

2008 AUTHORIZATION ACT

(a) In General- In order to stimulate commercial use of space, help maximize the utility and productivity of the International Space Station, and enable a commercial means of providing crew transfer and crew rescue services for the International Space Station, NASA shall--

(1) make use of United States commercially provided International Space Station crew transfer and crew rescue services to the maximum extent practicable, if those commercial services have demonstrated the capability to meet NASA-specified ascent, entry, and International Space Station proximity operations safety requirements;

(2) limit, to the maximum extent practicable, the use of the Crew Exploration Vehicle to missions carrying astronauts beyond low Earth orbit once commercial crew transfer and crew rescue services that meet safety requirements become operational;

(3) facilitate, to the maximum extent practicable, the transfer of NASA-developed technologies to potential United States commercial crew transfer and rescue service providers, consistent with United States law; and

(4) issue a notice of intent, not later than 180 days after the date of enactment of this Act, to enter into a funded, competitively awarded Space Act Agreement with 2 or more commercial entities for a Phase 1 Commercial Orbital Transportation Services crewed vehicle demonstration program."

2008 APPROPRIATIONS ACT

"...encourage(d) NASA to consider exercising its option for the Commercial Cargo Capability (COTS) Capability D (crew transport) as soon as possible..."

NATIONAL SECURITY PRESIDENTIAL DIRECTIVE - 49

"...departments and agencies shall use commercial space capabilities and services to the maximum practical extent."

Source: U.S. Government

Figure iii. Congressional guidance in FY 2008 NASA Authorization and Appropriation Acts and other national policies concerning commercial use of space and commercial crew capabilities

In the 1920's the federal government also supported the growth of air transportation by investing in technology. The Committee strongly believes it is time for NASA to reassume its crucial role of developing new technologies for space. Today, the alternatives available for exploration systems are severely limited because of the lack of a strategic investment in technology development in past decades. NASA now has an opportunity to generate a technology roadmap that aligns with an exploration mission that will last for decades. If appropriately funded, a technology development program would re-engage minds at American universities, in industry, and within NASA. The investments should be designed to increase the capabilities and reduce the costs of future exploration. This will benefit human and robotic exploration, the commercial space community, and other U.S. government users alike.

Future Destinations for Exploration

What is the strategy for exploration beyond low-Earth orbit? Humans could embark on many paths to explore the inner solar system, most particularly the following:

- Mars First, with a Mars landing, perhaps after a brief test of equipment and procedures on the Moon.
- Moon First, with lunar surface exploration focused on developing the capability to explore Mars.
- A Flexible Path to inner solar system locations, such as lunar orbit, Lagrange points, near-Earth objects and the moons of Mars, followed by exploration of the lunar surface and/or Martian surface.

A human landing followed by an extended human presence on Mars stands prominently above all other opportunities for exploration. Mars is unquestionably the most scientifically interesting destination in the inner solar system, with a planetary history much like Earth's. It possesses resources that can be used for life support and propellants. If humans are ever to live for long periods on another planetary surface, it is likely to be on Mars. But Mars is not an easy place to visit with existing technology and without a substantial investment of resources. The Committee finds that Mars is the ultimate destination for human exploration of the inner solar system, but it is not the *best* first destination.

What about the Moon first, then Mars? By first exploring the Moon, we could develop the operational skills and technology for landing on, launching from and working on a planetary surface. In the process, we could acquire an understanding of human adaptation to another world that would one day allow us to go to Mars. There are two main strategies for exploring the Moon. Both begin with a few short sorties to various sites to scout the region and validate lunar landing and ascent systems. In one strategy, the next step would be to build a lunar base. Over many missions, a small colony of habitats would be assembled, and explorers would begin to live there for many months, conducting scientific studies and prospecting for resources to use as fuel. In the other strategy, sorties would continue to different sites, spending weeks and then months at each one. More equipment would have to be brought to the lunar surface on each trip, but more diverse sites would be explored and in greater detail.

There is a third possible path for human exploration beyond low-Earth orbit, which the Committee calls the Flexible Path. On this path, humans would visit sites never visited before and extend our knowledge of how to operate in space—while traveling greater and greater distances from Earth. Successive missions would visit lunar orbit; the Lagrange points (special points in space that are important sites for scientific observations and the future space transportation infrastructure); and near-Earth objects (asteroids and spent comets that cross the Earth's path); and orbit around Mars. Most interestingly, humans could rendezvous with a moon of Mars, then coordinate with or control robots on the Martian surface, taking advantage of the relatively short communication times. At least initially, astronauts would not travel into the deep gravity wells of the lunar and Martian surface, deferring the cost of developing human landing and surface systems.

The Flexible Path represents a different type of exploration strategy. We would learn how to live and work in space, to visit small bodies, and to work with robotic probes on the planetary surface. It would provide the public and other stakeholders with a series of interesting "firsts" to keep them engaged and supportive. Most important, because the path is flexible, it would allow for many different options as exploration progresses, including a return to the Moon's surface or a continuation directly to the surface of Mars.

The Committee finds that both Moon First and Flexible Path are viable exploration strategies. It also finds that they are not necessarily mutually exclusive; before traveling to Mars, we might be well served to both extend our presence in free space and gain experience working on the lunar surface.

Integrated Program Options

The Committee has identified five principal alternatives for the human spaceflight program. They include one baseline case, which the Committee considers to be an executable version of the current program of record, funded to achieve its stated exploration goals, as well as four alternatives. These options and several derivatives are summarized in Figure iv.

The Committee was asked to provide two options that fit within the FY 2010 budget profile. This funding is essentially flat or decreasing through 2014, then increases at 1.4 percent per year thereafter, less than the 2.4 percent per year used by the Committee to estimate cost inflation. The first two options are constrained to the existing budget.

	Budget	Shuttle Life	ISS Life	Heavy Launch	Crew to LEO
Constrained Options					
Option 1: Program of Record (constrained)	FY10 Budget	2011	2015	Ares V	Ares I + Orion
Option 2: ISS + Lunar (constrained)	FY10 Budget	2011	2020	Ares V Lite	Commercial
Moon First Options					
Option 3: Baseline - Program of Record	Less constrained	2011	2015	Ares V	Ares I + Orion
Option 4A: Moon First - Ares Lite	Less constrained	2011	2020	Ares V Lite	Commercial
Option 4B: Moon First - Extend Shuttle	Less constrained	2015	2020	Directly Shuttle Derived + refueling	Commercial
Flexible Path Options					
Option 5A: Flexible Path - Ares Lite	Less constrained	2011	2020	Ares V Lite	Commercial
Option 5B:Flexible Path - EELV Heritage	Less constrained	2011	2020	75mt EELV + refueling	Commercial
Option 5C: Flexible Path - Shuttle Derived	Less constrained	2011	2020	Directly Shuttle Derived + refueling	Commercial

Source: Review of U.S. Human Spaceflight Plans Committee 1

Figure iv. A summary of the Integrated Options evaluated by the Committee

Option 1. Program of Record as assessed by the committee, constrained to the FY 2010 budget

This option is the program of record, with only two changes the Committee deems necessary: providing funds for the Shuttle into FY 2011 and including sufficient funds to de-orbit the ISS in 2016. When constrained to this budget profile, Ares I and Orion are not available until after the ISS has been de-orbited. The heavy-lift vehicle, Ares V, is not available until the late 2020s, and there are insufficient funds to develop the lunar lander and lunar surface systems until well into the 2030s, if ever.

Option 2. ISS and Lunar Exploration, constrained to FY 2010 Budget

This option extends the ISS to 2020, and begins a program of lunar exploration using a derivative of Ares V, referred to here as the Ares V Lite. The option assumes completion of the Shuttle manifest in FY 2011, and it includes a technology development program, a program to develop commercial services to transport crew to low-Earth orbit, and funds for enhanced utilization of the ISS. This option does not deliver heavy-lift capability until the late 2020s and does not have funds to develop the systems needed to land on or explore the Moon in the next two decades.

The remaining three alternatives fit a different budget profile—one that the Committee judged more appropriate for an exploration program designed to carry humans beyond low-Earth orbit. This budget increases to $3 billion above the FY 2010 guidance by FY 2014, then grows with inflation at what the Committee assumes to be 2.4 percent per year.

Option 3. Baseline case—Implementable program of record. this is an executable version of the program of record

It consists of the content and sequence of that program–de-orbiting the ISS in 2016, developing Orion, Ares I and Ares V, and beginning exploration of the Moon using the Altair lander and lunar surface systems. The Committee made only two additions it felt essential: budgeting for the completion of remaining flights on the Shuttle manifest in 2011 and including additional funds for the de-orbit of the ISS. The Committee's assessment is that, under this funding profile, the option delivers Ares I and Orion in FY 2017, with human lunar return in the mid-2020s.

Option 4. Moon first

This option preserves the Moon as the first destination for human exploration beyond low- Earth orbit. It also extends the ISS to 2020, funds technology advancement, and uses commercial vehicles to carry crew to low-Earth orbit. There are two significantly different variants to this option. Both develop the Orion, the Altair lander and lunar surface systems as in the Baseline Case.

Variant 4A is the Ares V Lite variant. This option retires the Shuttle in FY 2011 and develops the Ares V Lite heavy-lift launcher for lunar exploration. *Variant 4B* is the Shuttle extension variant. It offers the only foreseeable way to eliminate the gap in U.S. human-launch capability: by extending the Shuttle to 2015 at a minimum safe-flight rate. It also takes advantage of synergy with the Shuttle by developing a heavy-lift vehicle that is more directly

Shuttle-derived than the Ares family of vehicles. Both variants of Option 4 permit human lunar return by the mid-2020s.

Option 5. Flexible path

This option follows the Flexible Path as an exploration strategy. It operates the Shuttle into FY 2011, extends the ISS until 2020, funds technology advancement and develops commercial services to transport crew to low-Earth orbit. There are three variants within this option. They all use the Orion crew exploration vehicle, together with new in-space habitats and propulsions systems. The variants differ only in the heavy-lift vehicle used.

Variant 5A is the Ares V Lite variant. It develops the Ares V Lite, the most capable of the heavy-lift vehicles in this option. *Variant 5B* employs an EELV-heritage commercial heavy-lift launcher and assumes a different (and significantly reduced) role for NASA. It has an advantage of potentially lower operational costs, but requires significant restructuring of NASA. *Variant 5C* uses a Shuttle-derived, heavy-lift vehicle, taking maximum advantage of existing infrastructure, facilities and production capabilities.

All variants of Option 5 begin exploration along the flexible path in the early 2020s, with lunar fly-bys, visits to Lagrange points and near-Earth objects and Mars fly-bys occurring at a rate of about one major event per year, and possible rendezvous with Mars's moons or human lunar return by the mid- to late-2020s.

The Committee has found two executable options that comply with the FY 2010 budget profile. However, neither allows for a viable exploration program. In fact, the Committee finds that no plan compatible with the FY 2010 budget profile permits human exploration to continue in any meaningful way.

The Committee further finds that it is possible to conduct a viable exploration program with a budget rising to about $3 billion annually in real purchasing power above the FY 2010 budget profile. At this budget level, both the Moon First and the Flexible Path strategies begin human exploration on a reasonable but not aggressive timetable. The Committee believes an exploration program that will be a source of pride for the nation requires resources at such a level.

Organizational and Programmatic Issues

How might NASA organize to explore? The NASA Administrator needs to be given the authority to manage NASA's resources, including its workforce and facilities. It is noted that even the best-managed human spaceflight programs will encounter developmental problems. Such activities must be adequately funded, including reserves to account for the unforeseen and unforeseeable. Good management is especially difficult when funds cannot be moved from one human spaceflight budget line to another—and where additional funds can ordinarily be obtained only after a two-year delay (if at all). NASA would become a more effective organization if it were given the flexibility possible under the law to establish and manage its programs.

Finally, significant space achievements require continuity of support over many years. Program changes should be made based on future costs and future benefits and then only for compelling reasons. NASA and its human spaceflight program are in need of stability in both resources and direction. This report of course offers options that represent changes to the present program—along with the pros and cons of those possible changes. It is necessarily left to the decision-maker to determine whether these changes rise to the threshold of "compelling."

Summary of Principal Findings

The Committee summarizes its principal findings below. Additional findings are included in the body of the report.

The right mission and the right size

NASA's budget should match its mission and goals. Further, NASA should be given the ability to shape its organization and infrastructure accordingly, while maintaining facilities deemed to be of national importance.

International partnerships

The U.S. can lead a bold new international effort in the human exploration of space. If international partners are actively engaged, including on the "critical path" to success, there could be substantial benefits to foreign relations and more overall resources could become available to the human spaceflight program.

Short-term Space Shuttle planning

The remaining Shuttle manifest should be flown in a safe and prudent manner without undue schedule pressure. This manifest will likely extend operation into the second quarter of FY 2011. It is important to budget for this likelihood.

The human-spaceflight gap

Under current conditions, the gap in U.S. ability to launch astronauts into space will stretch to at least seven years. The Committee did not identify any credible approach employing new capabilities that could shorten the gap to less than six years. The only way to significantly close the gap is to extend the life of the Shuttle Program.

Extending the International Space Station

The return on investment to both the United States and our international partners would be significantly enhanced by an extension of the life of the ISS. A decision not to extend its operation would significantly impair U.S. ability to develop and lead future international spaceflight partnerships.

Heavy lift

A heavy-lift launch capability to low-Earth orbit, combined with the ability to inject heavy payloads away from the Earth, is beneficial to exploration. It will also be useful to the national security space and scientific communities. The Committee reviewed: the Ares family of launchers; Shuttle- derived vehicles; and launchers derived from the Evolved Expendable

Launch Vehicle family. Each approach has advantages and disadvantages, trading capability, life-cycle costs, maturity, operational complexity and the "way of doing business" within the program and NASA.

Commercial launch of crew to low-Earth orbit

Commercial services to deliver crew to low-Earth orbit are within reach. While this presents some risk, it could provide an earlier capability at lower initial and life-cycle costs than government could achieve. A new competition with adequate incentives to perform this service should be open to all U.S. aerospace companies. This would allow NASA to fo-cus on more challenging roles, including human exploration beyond low-Earth orbit based on the continued development of the current or modified Orion spacecraft.

Technology development

For exploration and commercial space: Investment in a well-designed and adequately funded space technology program is critical to enable progress in exploration. Exploration strategies can proceed more readily and economically if the requisite technology has been developed in advance. This investment will also benefit robotic exploration, the U.S. commercial space industry, the academic community and other U.S. government users.

Pathways to Mars

Mars is the ultimate destination for human exploration of the inner solar system; but it is not the best first destination. Visiting the "Moon First" and following the "Flexible Path" are both viable exploration strategies. The two are not necessarily mutually exclusive; before traveling to Mars, we could extend our presence in free space and gain experience working on the lunar surface.

Options for the human spaceflight program

The Committee developed five alternatives for the Human Spaceflight Program. It found:

- Human exploration beyond low-Earth orbit is not viable under the FY 2010 budget guideline.
- Meaningful human exploration is possible under a less- constrained budget, increasing annual expenditures by approximately $3 billion in real purchasing power above the FY 2010 guidance.
- Funding at the increased level would allow either an exploration program to explore the Moon First or one that follows the Flexible Path. Either could produce significant results in a reasonable timeframe.

1. Introduction

The Executive Office of the President established the Review of U.S. Human Spaceflight Plans Committee to develop options "in support of planning for U.S. human spaceflight activities beyond the retirement of the Space Shuttle." The Committee was asked to review the program of record and offer prospective alternatives, not to recommend a specific future course for the human spaceflight program. The Committee consisted of 10 individuals versed in the history, challenges and existing policies and plans for human spaceflight, members representing a broad and diverse set of views on spaceflight's possible future. The Committee's deliberations in its seven public sessions were informed by dozens of briefings, several site visits, and hundreds of documents received directly or through its website.

The current U.S. human spaceflight program appears to be on an unsustainable trajectory. It is perpetuating the perilous practice of pursuing goals that are often admirable, but which do not match available resources. President Kennedy stated, "We choose to ... do [these] things, not because they are easy, but because they are hard..." And, indeed, space operations are among the most complex and demanding activities ever undertaken by humans. It really is rocket science. Space operations become all the more difficult when means do not match aspirations. Such is the case today. The human spaceflight program, in the opinion of this Committee, is at a tipping point where either additional funds must be provided or the exploration program first instituted by President Kennedy must be abandoned at least for the time being.

America continues to enjoy a clear global leadership role in space capabilities. NASA's accomplishments are legion. Foremost among these is the landing of 12 astronauts on the Moon and returning them all safely to Earth. At that time, optimism was such that a study chaired by then-Vice President Agnew provided options to place humans on Mars by the mid-1980s—less than two decades after the initial lunar landing. (See Figure 1-1.)

But that was 40 years ago. The last person to stand on the Moon returned to Earth 37 years ago. (See Figure 1-2.) Since the end of the Apollo Program, no American has traveled more than 386 miles from the surface of the Earth. Some 70 percent of Americans living today had not yet been born at the time of Apollo 11.

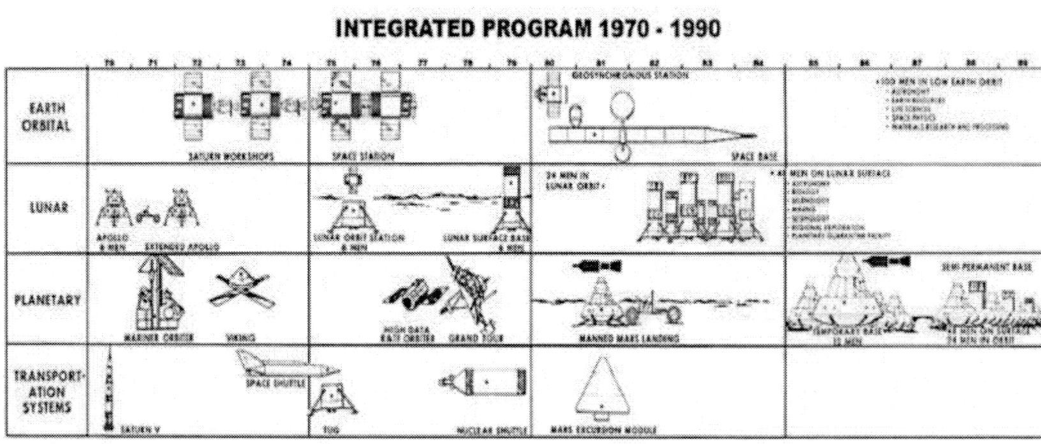

Source: NASA

Figure 1-1. The integrated program that never was. The human spaceflight program that was expected to follow the initial Apollo lunar missions. Only a space shuttle and space station have been developed so far

Source: NASA (Apollo 17)

Figure 1-2. Astronaut Gene Cernan as photographed by astronaut Jack Schmitt on the sixth and final Apollo exploration of the lunar surface in 1972

Today, the nation faces important decisions about the future of human spaceflight. Will we again leave the close proximity of low-Earth orbit and explore the solar system, charting a path for the eventual expansion of human civilization into space? If so, how will we ensure that our exploration delivers the greatest benefit to the nation? Can we explore with reasonable assurance of human safety? And can the nation marshal the resources to embark on the mission? Although there remain significant potential barriers to prolonged deep- space operations, which deserve greater attention than they are currently receiving (e.g., adaptation of humans to the micro-gravity and radiation environments of space away from the protective features of the Earth), the principal barrier to space operations continues to be its high cost compared with the resources that have been available.

Space exploration, initially a competitive pursuit, has become a global enterprise. Many other nations have aspirations in space, and the combined annual budgets of their space programs are comparable to NASA's. If the U.S. is willing to lead a global program of exploration, sharing both the burdens and benefits of space exploration in a meaningful way, significant benefits could follow. Actively engaging international partners in a manner adapted to today's multipolar world could strengthen geopolitical relationships, leverage global financial and technological resources, and enhance the exploration enterprise.

In addition, there is now a burgeoning commercial space in-dustry. Given the appropriate incentives, this industry might help overcome a long-standing problem. The cost of admission to a variety of space activities strongly depends on the cost of reaching low-Earth orbit. These costs become even greater when, as is the circumstance today, large sums are paid to develop new launch systems but those systems are used only infrequently. It seems improbable that order-ofmagnitude reductions in launch costs will be realized until launch rates increase substantially. But this is a "chickenand-egg" problem. The early airlines faced a similar barrier, which was finally resolved when the federal government awarded a series of guaranteed contracts for carrying the mail. A corresponding action may be required if space is ever to become broadly accessible. If we craft a space architecture to provide opportunities to industry, creating an assured initial market, there is the potential—not without risk—that the eventual costs to the government could be reduced substantially.

Significantly, we are more experienced than we were in 1961, and we are able to build on that experience as we design an exploration program. If, after designing cleverly, building alliances with partners, and engaging commercial providers, the nation cannot afford to fund the effort to pursue the goals it would like to embrace, it should accept the disappointment of setting lesser goals. Whatever space program is ultimately selected, it must be matched with the resources needed for its execution. Here lies NASA's greatest peril of the past, present, and—absent decisive action—future. These challenging initiatives must be adequately funded, including reserves to account for the unforeseen and unforeseeable. (See Figure 1-3.)

Can we explore with reasonable assurance of human safety? Human space travel has many benefits, but it is an inherently dangerous endeavor. Past gains in launch systems reliability and safety have been realized at a painfully slow pace. Predictive models have generally proven unsatisfactory in accurately forecasting absolute reliability—many actual failures have been attributable to causes not included in most reliability models (e.g., process errors, design flaws, and, less frequently, operational errors). A great deal has been learned in building more reliable space systems, and this is not to suggest otherwise; rather, it is to confirm that this is an area deserving continuing attention. Human safety can never be absolutely assured, but throughout this report, safety is treated as the *sine qua non*. Concepts falling short in human safety have simply been eliminated from consideration. For example, no options proceeding *directly* to Mars have been offered as alternatives, because the Committee believes the state of technology, the understanding of risks, and the available operational experience are sufficiently immature—irrespective of the budgetary limitations—to commit to such an endeavor.

	Real Year Dollars (Billions)	2009 Constant Dollars (Billions) Using GDP Deflator
Mercury (1959-1963)	0.3	1.6
Gemini (1962-1967)	1.3	7.2
Apollo (1961-1973)	24.6	129.5
Shuttle (1971-2009)	112.8	172.5
ISS (1994-2009)	31.5	35.2
Constellation (2006-2020)	108.2	98.4

Source: NASA

Notes:

1. Mercury, Gemini, Apollo, Shuttle, and ISS costs are actual costs derived from historical budget documents.
2. Constellation costs are estimates that are supplied by the Constellation Program Office and based on an unconstrained budget that cumulates in a single Human Lunar return mission in 2020.

Figure 1-3. Human Spaceflight Programs Costs in Real Year and Constant Year 2009 Dollars

How will we explore to deliver the greatest benefit to the nation? Planning for a human spaceflight program should begin with a choice of goals—rather than a choice of destinations. Destinations should derive from goals, and alternative architectures may be weighed against those goals. There is now a strong consensus in the United States that the next step in human spaceflight should be to travel beyond low-Earth orbit. This promises to provide important benefits to society, including driving technological innovation; developing commercial industries and important national capabilities; and contributing to our expertise in further exploration. Human exploration *can* contribute appropriately to the expansion of scientific knowledge, especially field geology, and it is in the interest of both science and human spaceflight that a credible and well-rationalized strategy of coordination between the two endeavors be developed. Robotic spacecraft will play an important role as a precursor to human spaceflight activi ties. The Committee concluded that the ultimate goal of human exploration is to chart a path for human expansion into the solar system. This is an ambitious goal, but one worthy of U.S. leadership in concert with a broad range of international partners.

With regard to the human spaceflight program itself, the Committee has been deluged with strongly and genuinely held, frequently conflicting, beliefs as to the program's proper composition. For example, the following statements appeared in six different communications that happened to come across the Committee Chairman's desk within minutes of each other:

- "As an American, having NASA field a retro-reenactment of the Apollo program to get back to the moon a half-century after we sent people there the first time is humiliating."
- "From a safety and continuity standpoint the next step in space must be a return to the moon."
- "I am an aerospace engineering master's candidate. [My classmates'] options are working for monolithic bureaucracies where their creativity will be crushed by program cancellations, cost overruns and risk aversion... It is no surprise that many of them choose to work in finance..."
- "We remember the past well and remind ourselves often of long gone civilizations whose innovations in science, technology and learning yielded knowledge that served as beacons of brilliance, but who lost the spark and faded."
- "...going back to the moon takes us into an intellectual and political cul de sac..."
- "The audacity to go to the moon was perhaps the 20th century's greatest illustration of America's optimism. Present generations of Americans need to capture some of that audacity."

A primary issue in formulating a human spaceflight plan is its affordability. In the way of background, Figures 1-4, 1-5 and 1-6 present the overall NASA budget trend over time in absolute terms and in relationship to the GDP and the federal budget, respectively. The trend in funding the human spaceflight portion of NASA's portfolio is shown in Figure 1-7. Today, the human spaceflight program costs each citizen about seven cents a day.

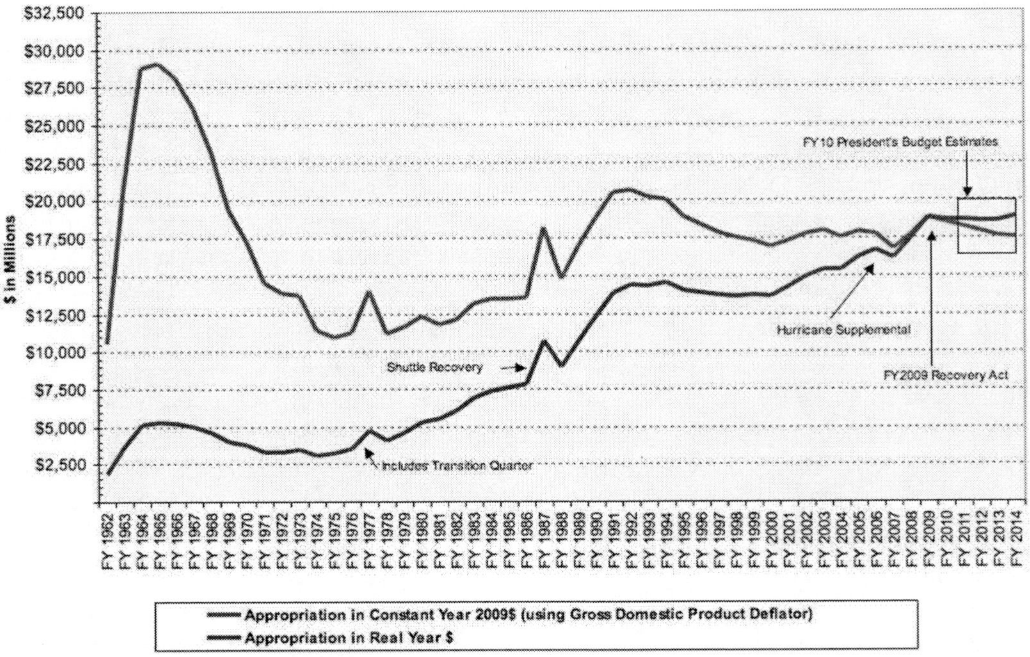

Source: OMB Historical Budget Tables

Figure 1-4. NASA Appropriation History in Real Year and Constant Year 2009 Dollars

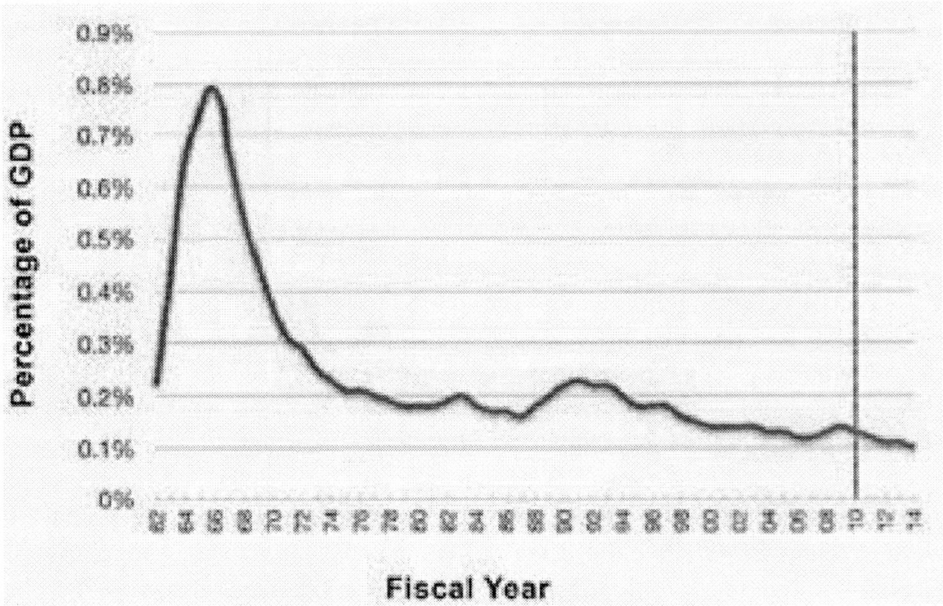

Source: OMB Historical Budget Tables

Figure 1-5. As a percent of Gross Domestic Product the NASA budget has more or less continuously diminished since the peak of the Apollo program

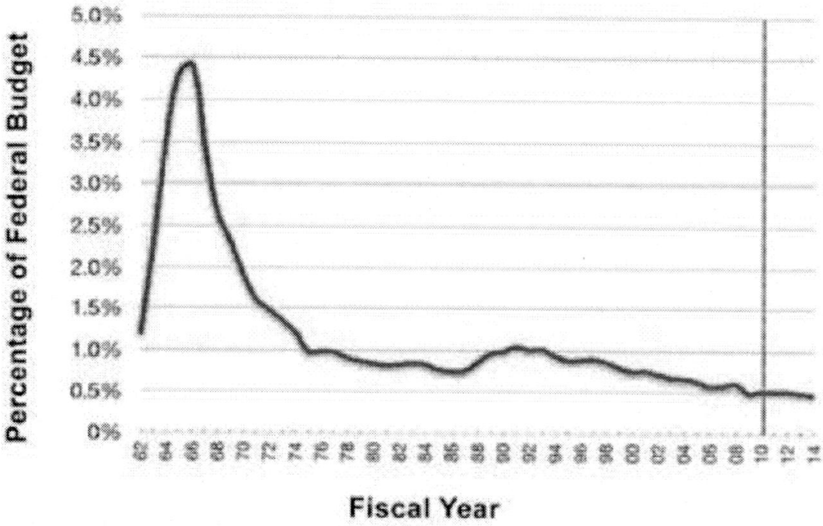

Source: OMB Historical Budget Tables

Figure 1-6. The overall NASA budget as a fraction of the federal budget has declined from 4.5 percent at the peak of the Apollo program to approximately 0.5 percent today.

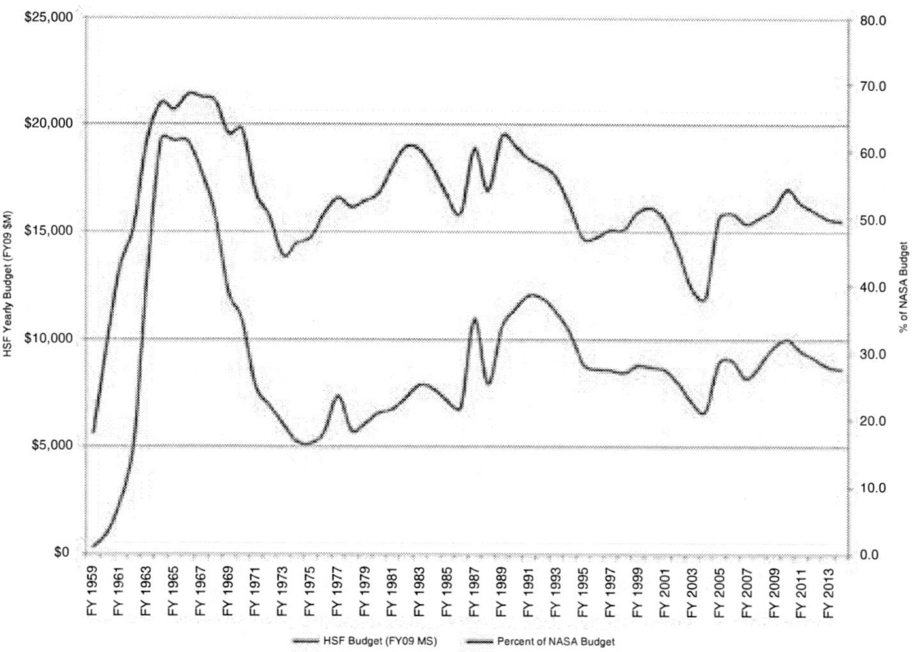

Source: NASA

Figure 1-7. Human spaceflight yearly annual budget in FY 2009 dollars (left scale) and as a percentage of total NASA budget (right scale.)

So what should America's human spaceflight program look like? Before answering that question, we must face an underlying reality. We are where we are. The Committee thus identified five questions that could form the basis of a plan for U.S. human spaceflight:

- What should be the future of the Space Shuttle?
- What should be the future of the International Space Station?
- On what should the next heavy-lift launch vehicle be based?
- How should crew be carried to low-Earth orbit?
- What is the most practicable strategy for exploration beyond low-Earth orbit?

The Committee considers the framing of these questions, in a consistent way, to be at least as important as their combinations in the integrated options for a human spaceflight plan. The Committee assessed the programs within the current human spaceflight portfolio, considered capabilities and technologies that a future program might require, and examined the roles of commercial industry and our international partners in this enterprise.

A human landing and extended human presence on Mars stand prominently above all other opportunities for exploration. (See Figure 1-8.) Mars is unquestionably the most scientifically interesting destination in the inner solar system. It possesses resources which can be used for life support and propellants. If humans are ever to live for long periods with intention of extended settlement on another planetary surface, it is likely to be on Mars. But Mars is not an easy place to visit with existing technology and without a substantial investment in resources. The Committee concluded that Mars is the ultimate destination for human exploration of the inner solar system; but as already noted, it is not the best first destination.

The Committee thus addressed several possible strategies for exploration beyond low-Earth orbit. We could choose to explore the Moon first, with lunar surface exploration focused on developing the capability to explore Mars. Or we could choose to follow a flexible path to successively distant or challenging destinations, such as lunar orbit, Lagrange points, near-Earth objects, or the moons of Mars, which could lead to the possible exploration of the lunar surface and/or Martian surface.

Source: NASA

Figure 1-8. Artist's concept of Mars mission activity

As a result of its deliberations, the Committee developed five integrated options for the U.S. human spaceflight program that the Committee deems representative: one baseline case, founded upon the Constellation program, and four alternatives. Two of the options are constrained to the FY 2010 budget profile. The remaining three options, including the baseline, fit a less-constrained budget. It was possible to define some 3,000 potential options from the set of parameters considered—hence the options presented here should be thought of as representative families. Various program additions and deletions among these families are also plausible, with appropriate budget adjustments—including a proper accounting of the many interdependent facets of these integrated options. Several of these derivatives are discussed in this report.

The Committee considers it important for any exploration strategy to offer a spectrum of choices that provides periodic milestone accomplishments as well as a continuum of investment cost options. Unfortunately, for all options examined, the "entry cost" for human exploration is indeed significant—and for the more inspiring options there does not seem to be a "cost continuum." Put another way, there is a sizeable difference in the cost of programs between those operating in low-Earth orbit and those exploring beyond low-Earth orbit.

Clearly, a more penetrating analysis into any choice will be required before fully embarking upon it. However, the Committee believes it has fairly represented the most plausible courses. It bases this assessment in part on the extraordinary supporting effort provided by NASA personnel—an effort that was forthright, competent, and, in the NASA spirit, "can-do." The Committee also benefited significantly from prior independent reviews of NASA activities. In addition, the Committee contracted with the Aerospace Corporation to provide independent assessments. During the Committee's deliberations, it was informed by day-long public meetings in Houston, TX, Huntsville, AL, and Cocoa Beach, FL, as well as five days of meetings in Washington, DC. In addition, its subcommittees held meetings in Denver, CO; Decatur, AL; Huntsville, AL; Michoud, LA; Hawthorne, CA; El Segundo, CA; and Dulles, VA. The group conducted numerous teleconference and videoconference preparatory sessions and communicated frequently by e-mail (over 1,700 e-mails in the Chairman's case).

Seeking to benefit from the views of the public, the Committee: established a website and Facebook site; used Twitter; conducted all decisional meetings in public session (meetings that were also carried on NASA TV); provided opportunity for public comment at five of the formal meetings; testified before committees of both the House and Senate; and held seven press conferences. Participation by the public was extensive—and the Committee made use of that input. It is heartening to note that the public still strongly supports the overall efforts of NASA. (See Figure 1-9.)

Finally, it is reasonable to ask whether a review over several months is sufficient to offer the options presented here. Certainly, the issues at hand demand a broad and detailed understanding of the human spaceflight program—ranging from an awareness of the impact of galactic cosmic rays on the human body to the fact that the hook-height at NASA's Michoud Assembly Facility will only allow the manufacture of a stage with a diameter of 33 feet.

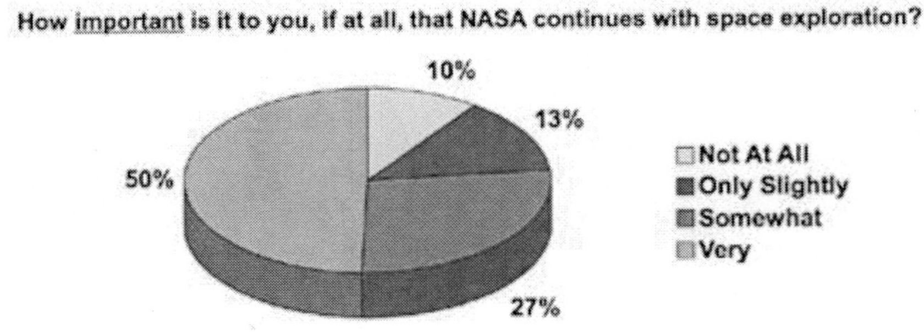

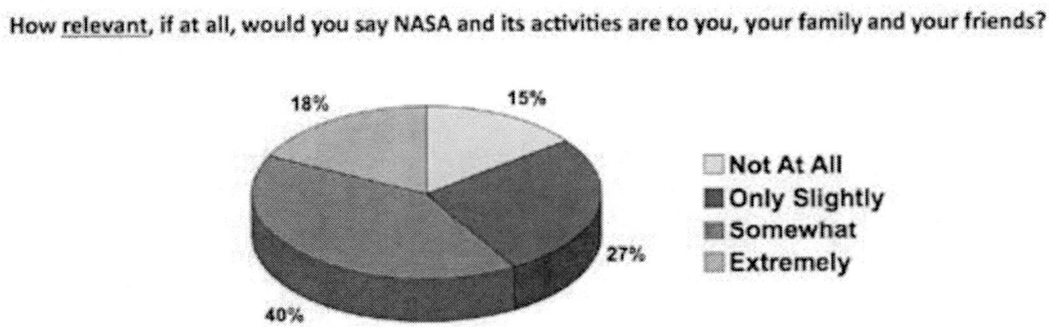

Source: NASA - 2009 Market Research Insights and Implications - August 6, 2009

Figure 1-9. Space exploration (human/robotic) continues to be valued by Americans – 77% think it is very or somewhat important, while 58% think it extremely or somewhat relevant.

Each of the Committee members had accrued extensive experience with spaceflight issues long before the beginning of this review. For example, the members cumulatively have amassed 245 days in orbit, 6 flights into space, 293 years working on space matters, 175 years in science, 144 years in engineering, 143 years in engineering management, 61 years in space operations, 77 years in government, 35 years in the military, and 160 years in the private sector. (The totals reflect the overlap of some of these categories.)

The Committee believes that the options presented here, if matched with appropriate funds, provide a reasonable foundation for selecting a human spaceflight program worthy of a great nation

2. U.S. Human Spaceflight: Historical Review

In human spaceflight, as in other endeavors, a review of the past can help provide perspective in planning for the future. This chapter, noting the findings of some earlier space-program assessments, seeks to provide such perspective.

In 1961, President John F. Kennedy forcefully and publicly focused the nation's nascent space program on a single goal: U.S. astronauts would set foot on the surface of the Moon before the end of the decade, and return safely. More feasible than a mission to Mars, and with better prospect of preceding the Soviets than seeking to develop an orbiting space station, the lunar landing would ensure U.S. standing as the leader in the world's most prominent exploration competition. Kennedy's challenge to NASA and the nation was an audacious one, given that at the time he made it in 1961, no American had even reached Earth orbit.

President Kennedy markedly accelerated the U.S. space program, but he did not ini-tiate it. NASA was established in 1958. President Dwight Eisenhower supported human missions to low-Earth orbit and beyond, but he emphasized fiscal restraint in the effort. According to George Low, at the time NASA Chief of Manned Spaceflight, a desirable lunar program should have project costs kept in balance with expected returns and within the foreseeable NASA budget. By the 1960s, the drive to meet the end-of-decade goal superseded those restraints.

The outcome of the so-called Space Race was not a foregone conclusion. Among numerous Soviet achievements, it was a cosmonaut who was the first human to orbit the Earth, and another to first conduct extra-vehicular activity. The U.S. Mercury and Gemini programs achieved their objectives of developing and flight-testing the kinds of equipment and procedures that would be needed in a lunar mission. But in 1967, with less than three years remaining before the deadline set by President Kennedy, fire broke out in the pure-oxygen environment of the Apollo 1 command module during a ground test, resulting in the death of the three astronauts on board. Despite the delay from the accident and its aftermath, on July 20, 1969, Neil Armstrong and Buzz Aldrin of Apollo 11 became the first humans to set foot on a celestial body beyond our own, while Mike Collins orbited above, preparing for the return to Earth. Ten more astronauts, on five more Apollo missions, would reach the lunar surface.

The New York Times.

SOVIET FIRES EARTH SATELLITE INTO SPACE; IT IS CIRCLING THE GLOBE AT 18,000 M. P. H.; SPHERE TRACKED IN 4 CROSSINGS OVER U.S.

The New York Times

MEN WALK ON MOON

ASTRONAUTS LAND ON PLAIN; COLLECT ROCKS, PLANT FLAG

Source: The New York Times

Figure 2-1. Front page of The New York Times from October 5, 1957 and July 21, 1969

Big goals are energizing, for individuals and for national efforts. (See Figure 2-1) But the problem with focusing on a single all-consuming objective is the letdown that can ensue after the objective is achieved. Public interest appeared to wane during the course of succeeding lunar missions. NASA had ambitious set of plans to follow Apollo. Space stations would orbit the Earth. A more permanent lunar presence would be established by 1982. Proposed hardware included a space tug and a nuclear-powered shuttle. The first crewed landing on Mars would take place by the mid-1980s. By 1990, there would be 100 humans in low-Earth orbit, 48 on the Moon and 72 on Mars and its moons. Most of that never came to pass. Since 1972, the year of the last Apollo lunar mission, no human has ventured farther from the Earth's surface than 386 miles.

President Richard Nixon did not end the space program, but he did much to scale it back. The trajectory of the NASA budget shifted downward. The Nixon administration was responding not only to the perceived decline in public support for far-reaching human space exploration, but also to the economic decline at the time. When a task group established by the administration presented options that included a lunar return and a program aimed at Mars, the President confined the nation's crew-carrying space ventures instead to low- Earth orbit.

The keystone of the redefined initiative was the Space Shuttle, the reusable departure from the expendable transport systems used until that time, capable of launching as a rocket and landing as an aircraft. The economic case for the Shuttle was that it would provide dependable, high-frequency access to orbit, with relatively low cost. Government payloads, both civil and military, would be delivered aboard the orbiter, as well as commercial satellites. On some missions, the vehicle would serve as an orbiting laboratory. Despite its unprecedented technical complexity, the Shuttle's development budget was constrained, eliminating design options such as a fully reusable, two-stage configuration.

The first Space Shuttle reached orbit in April 1981, a little less than a decade after President Nixon announced the program in 1972. Launch frequency never approached original expectations, and the cost per mission turned out to be far greater than what was forecast. The plan to launch on nearly a weekly basis was cut to 24 flights per year, and even that proved unattainable. By January of 1986, five orbiters had flown four test flights and 20 operational missions.

On the morning of January 28, 1986, Space Shuttle *Challenger* was destroyed in an explosion 73 seconds after launch. The accident claimed the lives of all seven crew members. The Presidential Commission that investigated the accident, chaired by William Rogers, called for measures to correct critical design flaws in the Shuttle, as well as to correct management shortcomings it identified at NASA. The Commission advocated reasonable expectations for the Shuttle Program, urging that the space agency "establish a flight rate that is consistent with its resources."

After the Shuttle had been flying a variety of missions for a number of years, its primary purpose evolved to constructing and supporting space stations. Both the U.S. and the Soviet Union began operating orbiting platforms for research and other functions in the early 1970s. In his 1984 State of the Union Address, President Ronald Reagan announced plans to construct what became known as Space Station *Freedom*. James Beggs, NASA Administrator at the time, called this permanent orbiting facility "the next logical step" in space exploration. For President Reagan, the international project served as an element of foreign policy, helping to reinforce ties with allied nations. Later, after the end of the Soviet Union but before any joint space station was built, the station concept was promoted as a means to help foster cooperation with a former adversary.

In fact, *Freedom* was never built. The decade following the Reagan announcement saw a long series of design studies, redesigns and cost reassessments. Eventually, the original initiative, which included the U.S., Europe, Japan, and Canada, was expanded by joining forces with Russia to build the International Space Station. Before construction began on the International Space Station, the U.S. was building experience with the Russians, flying Shuttle missions to the Mir space station and flying NASA astronauts on Soyuz vehicles to long-duration Mir missions.

What tasks should a space station perform? Long-term, what should the United States be seeking to accomplish in space? In 1985, the sense in Congress was that it was not getting an adequate response from NASA and the White House, so, through legislation, it directed the establishment of an independent commission to examine these questions. Former NASA Administrator Thomas Paine chaired the National Commission on Space, which developed a half-century roadmap for the U.S. civil space program. Among numerous recommendations, the Commission counseled against focusing efforts on a single objective, on the Apollo model, with nothing to follow. It stressed program continuity, so there would not be another gap like the one between the end of Apollo and the beginning of the Shuttle. And it departed from the policy that had prevailed over the previous decade, limiting operational focus to low-Earth orbit. Humans were to return to the Moon by 2005 and reach Mars by 2015. The impact of the Paine Commission Report was diminished by timing: The *Challenger* accident occurred during the course of the Commission's inquiry.

In 1989, on the 20th anniversary of the Apollo 11 Moon landing, President George H. W. Bush announced the Space Exploration Initiative, which supported a number of the objectives spelled out in the earlier Paine Report, such as missions to the Moon and Mars. In the same speech, the President asked Vice President Dan Quayle to lead a National Space Council, which would determine the requirements to fulfill the Initiative. NASA Administrator Richard Truly, in turn, established a task force to support that inquiry. Among the findings of this "90-Day Study," the projected total cost of the proposed lunar and Mars projects, over 34 years, would be an estimated $541 billion (in 1991 dollars). In 1990, Congress zeroed the budget of the Space Exploration Initiative.

President Bush, Vice President Quayle and the Space Council called for a fresh assessment of the long-term prospects of NASA and the U.S. civil space program. To provide that assessment, an advisory committee was established, chaired by Norman R. Augustine, which raised numerous issues, starting with the lack of a national consensus on space program goals. Within NASA, the committee found an overextended agency, with shortcomings in budget, project development, personnel practices and other areas of management, and the committee cited the need for a heavy-lift launch vehicle and a space program balanced between human and robotic flight. The committee said the U.S. civil space program is "overly dependent upon the Space Shuttle for access to space." The committee also stated that "the statistical evidence indicates that we are likely to lose another Space Shuttle in the next several years." Among its prescriptions for improvement, the committee presented a new approach for long-range planning of space-exploration projects, in which programs would be "tailored to respond to the availability of funding, rather than adhering to a rigid schedule" that failed to recognize the impact of funding changes.

Twelve years later, on February 1, 2003, the nation did lose another Shuttle. *Columbia*, the first orbiter to reach space 22 years earlier, was destroyed during reentry, with the loss of all seven members of its crew. The Columbia Accident Investigation Board documented the physical cause of the accident, but also cited organizational and communications failures within NASA that allowed the critical damage to occur and go unaddressed. The report went on to cite "a lack, over the past three decades, of any national mandate providing NASA a compelling mission requiring human presence in space . . ."

Throughout NASA's history, while human spaceflight efforts garnered the most national attention, the agency continued to launch satellites, deep-space probes and rovers of ever-greater sophistication. The success of robotic missions such as the *Voyager* spacecraft to the outer planets fostered debate over the relative value of robotic versus human space exploration. In 1999, NASA Administrator Dan Goldin chartered a small internal task force—the Decadal Planning Team, which later evolved into a larger, agency-wide team known as the NASA Exploration Team—to investigate the best ways to coordinate human and robotic missions. These teams followed a series of architecture studies over the previous decade, such as the report of General Thomas Stafford and the Synthesis Group, all aimed at charting a renewed course of space exploration.

The work of the two NASA teams helped provide the basis for a new policy established by President George W. Bush in 2004, the Vision for Space Exploration. In announcing the Vision, the President acknowledged the numerous tangible benefits of space missions, in areas such as communications and weather forecasting, but the central purpose he stressed—as reflected in the name of the policy—was exploration, continuing the American tradition of discovery in uncharted territory. The new initiative echoed the earlier Eisenhower policy: fly well beyond Earth's realm, but do it on a fiscally sustainable basis.

Leading the agenda set out in the Vision was completion of the ISS by 2010. One reason cited was to meet the nation's obligations to its international partners; another was to investigate the effects on human biology of extended exposure to the space environment, thereby helping to develop the means to sustain astronauts on subsequent, long-duration missions. At least initially, the Vision did not stress the role of the ISS as a laboratory for other kinds of research. Completion of the ISS depended on returning the Space Shuttle to flight once safety concerns raised in the *Columbia* accident investigation were sufficiently addressed. Exploration beyond low-Earth orbit, under the Vision plan, was focused on the Moon—starting with robotic missions no later than 2008, followed by human return to the lunar surface by 2020. Once a human presence is well established on the Moon, the President said, "we will then be ready to take the next steps of space exploration: human missions to Mars and to worlds beyond."

At the time he announced the Vision, President Bush also appointed a commission, chaired by E. C. "Pete" Aldridge, Jr., to develop recommendations for implementing the plan. Among its recommendations, the commission said that NASA should "aggressively use its contractual authority to reach broadly into the commercial and nonprofit communities to bring the best ideas, technologies, and management tools into the accomplishment of exploration goals." Through its Commercial Orbital Transportation Services (COTS) program, NASA solicited proposals for private-sector transport of cargo and possibly crew to the International Space Station. Three awards were made, one of which was subsequently cancelled by NASA for failure to meet milestones.

In announcing the Vision, President Bush noted that "America has not developed a new vehicle to advance human exploration in space in nearly a quarter century." Proposals for a next-generation space vehicle had long been considered, but with the loss of *Columbia*, and with a mandate from the new Presidential policy to focus on the completion of the ISS, and then retire the Space Shuttle, NASA affirmatively began preparing for near-term retirement of the Shuttle. A newly constituted Exploration Systems Mission Directorate led the task of developing the Shuttle's successor. Initially, NASA chose a broad concept-maturation, risk reduction, and technology-investment approach to developing exploration systems.

In 2005, after Dr. Michael Griffin became Administrator, NASA undertook the Exploration Systems Architecture Study (ESAS) to select vehicles and systems in keeping with the Vision. The team evaluated hundreds of potential configurations. A leading objective was to minimize the gap between the last Shuttle flight and the first flight of the new vehicle. A date of 2012 was set for that first flight. Another criterion, spelled out in the 2005 Authorization Act for NASA, was to make use, as much as possible, of assets and infrastructure carried over from the Shuttle Program. Since the new system—actually a family of vehicles—would likely have a decades-long service life, it was to have the capability of not only reaching low-Earth orbit, but also extending to the Moon and beyond. The lunar objective would be to do more than replicate what Apollo had accomplished long before. The architecture would support larger crews and longer missions, capable of reaching any location on the Moon and returning. The results of the vehicle-system selection process evolved into what is now known as the Constellation Program, consisting of Ares launch vehicles, Orion crew capsule, the Altair lunar lander, and lunar surface systems.

Today, budget questions continue to dominate the human spaceflight debate. In the 37 years since humans last ventured beyond low-Earth orbit, and five years after announcement of the Vision for Space Exploration, consensus is still lacking about what is feasible and affordable in the future course of U.S. human spaceflight.

3. Goals and Future Destinations for Exploration

3.1. Goals for Exploration

We explore to reach goals, not destinations. It is in the definition of our goals that decision-making for human spaceflight should begin. With goals established, questions about destinations, exploration strategies and transportation architectures can follow in a logical order. While there are certainly some aspects of the transportation system that are common to all exploration missions (e.g. crew access and heavy lift to low-Earth orbit), there is a danger of choosing destinations and architectures first. This runs the risk of getting stuck at a destination without a clear understanding of why it was chosen, which in turn can lead to uncertainty about when it is time to move on.

Since 1972, the destination for U.S. human exploration of space has been confined to low-Earth orbit. Following the loss of *Columbia*, a strong national consensus emerged that we should move beyond low-Earth orbit once again, and explore the inner solar system. The question arises, "What is the point of doing so?" The answers to this question help to identify the goals of human spaceflight. While it was not specifically within the Statement of Task of the Committee to advise on the rationale for a human spaceflight program, the Committee felt compelled to at least review the likely goals as a foundation for its further deliberations.

Human spaceflight produces important tangible benefits to society. Human spaceflight is a technologically intensive activity, and during its execution new technologies are derived that have benefit to other government and commercial users of space, and to products that touch Americans daily. Access to and development of space is critical to our national welfare, and a well-crafted human exploration program can help to develop competitive commercial industries and important national capabilities. We explore our first destinations in part to learn how better to explore more challenging sites in the future. Human and robotic explorations both contribute to the expansion of scientific knowledge. Human explorers are most effective when exploring complex destinations, and particularly in endeavors such as field geology.

Human exploration also addresses larger goals. We live in an increasingly multi-polar world, and human space exploration is one domain in which the United States is still the acknowledged leader. Human exploration provides an opportunity to demonstrate space leadership while deeply engaging international partners. Chapter 8 will discuss the potential of partnerships in exploration.

Human exploration of space can engage the public in new ways, inspiring the next generation of scientists and engineers, and contributing to the development of the future workforce in science, technology, engineering and mathematics (STEM). By viewing other planets as well as our own from deep space, exploration helps to shape human perceptions of our place in the universe.

There was a strong consensus within the Committee that human exploration also should advance us as a civilization towards our ultimate goal: charting a path for human expansion into the solar system. It is too early to know how and when humans will first learn to live on another planet, but we should be guided by that long-term goal.

In developing alternatives for human spaceflight plans, the Committee was guided by these tangible, less-tangible and long-term goals. In Chapter 6, the Committee returns to these goals as the basis for developing evaluation measures against which the options will be evaluated.

3.2. Overview of Destinations and Approach

The Moon has been the nation's principal focus of human space exploration beyond low-Earth orbit since President Kennedy set it as a national goal in 1961. But there are many places humans could explore in the inner solar system, each with benefit to the public, as well as opportunities for scientific discoveries, technology development and steady progress in human exploration capabilities. Among these destinations are our own Moon, as well as Mars and its moons. (See Figure 3.2-1.) Other potential destinations include the near-Earth objects, asteroids and spent comets that pass near the Earth. There are also important locations in free space that are of interest, including the Earth's Lagrange points. These are sites at the edge of the Earth's influence, which will be important future points for observation toward the Earth and away from it. For example, the James Webb Space Telescope, the successor to the Hubble Space Telescope, will be placed at a Lagrange point. The Lagrange points might also be the nodes of a future space transportation highway through the inner solar system.

Source: Review of U.S. Human Spaceflight Plans Committee

Figure 3.2-1. Potential destinations for the U.S. human spaceflight program

There is a progression in time and difficulty in reaching these destinations. Lunar orbit, the Lagrange points, near-Earth objects, and a Mars fly-by are the easiest in terms of energy required. It actually requires less energy to fly by Mars than to land on and return from the surface of the Moon. Next in terms of energy requirements is the lunar surface, followed by Mars orbit. The surface of Mars requires the most energy to reach. An analysis of the duration to reach these destinations yields a slightly different order. The Moon is days away, the Lagrange points weeks, the near-Earth objects months, a Mars fly-by a year, and a Mars landing is the longest—about 900 days for a round trip using the most likely approach. Among practical criteria to apply in selecting destinations are questions such as: How difficult is the destination to reach? How long will it take? How dangerous will the mission be? How expensive and sustainable will it be?

The key framing question for this chapter is: What is the most practicable strategy for exploration beyond low-Earth orbit? Options include:

- Mars First, with a Mars landing after a brief test flight program of equipment and procedures on the Moon.
- Moon First, with surface exploration focused on developing capability for Mars.
- Flexible Path to Mars via the inner solar system objects and locations, with no immediate plan for surface exploration, then followed by exploration of the lunar and/or Martian surface.

In assessing these choices, the Committee examined a number of scenarios, performing new analyse and reviewing existing studies from the Constellation Program and other NASA architecture studies dating back to Apollo. The Committee listened carefully to alternate views presented by a number of other organizations, including the Mars Society and the Planetary Society.

With the help of a NASA team, the Committee examined the technical, programmatic and goal-fulfillment aspects of the destinations in two major cycles. The first cycle considered six destination/pathway scenarios, including three variants of lunar exploration, two variants of Mars exploration, and the "Flexible Path." (One scenario was the baseline Constellation Program approach.) For each of these six variants, benefits and timelines were developed, along with the necessary hardware elements and approximate costs. A second cycle included more detailed analysis, such as coupling to launch vehicles and upper stages, and more detailed benefit and cost analysis. In the end, the six initial scenarios collapsed into the three destination options described below.

Later, Chapter 6 provides a more complete description of each scenario, combining in each case the choice of destination with the choice of launch system to low-Earth orbit.

3.3. Mars First

3.3.1. Overview

A human landing that leads to an extended human presence on the Martian surface stands prominently above all other opportunities for human space exploration. Mars is somewhat smaller than Earth, has about three-eights its surface gravity, a thin atmosphere consisting mostly of carbon dioxide, and water. (See Figure 3.3.1-1.) It therefore possesses potential resources that can be used for life support and propellent. If humans are ever to live for long durations on another planetary surface and move toward permanent expansion of human civilization beyond the Earth, it is likely to be on Mars (Figure 3.3.1-2.) Mars is unquestionably the most scientifically interesting destination in the inner solar system. Mars has a planetary history similar to that of the Earth. It had a period of volcanic activity. At one time, water ran freely on its surface. Its atmosphere evolved over time, much as ours did. And there is the distinct potential that life could have begun to evolve on Mars. Learning about Mars would teach us a great deal about the Earth. Furthermore, the scientific community that studies Mars generally agrees that its exploration could be significantly enhanced by direct participation of astronaut explorers. The Committee finds that Mars is the ultimate destination for human exploration of the inner solar system.

3.3.2. Scenario Descriptions

Two scenarios have been developed to examine the human exploration of Mars. In the first, the surface of Mars would be the initial and only destination, and all resources would be focused on reaching it as soon as possible. In the second, systems would be designed for Mars missions, but would be first verified on several test flights to the Moon. The latter would require some hardware modification, but would test the systems at a planetary body near the Earth before committing to a multi-year mission to Mars. In the end, the Committee decided to use the variant with a brief test flight program of equipment and procedures on the Moon as the reference Mars First option.

	Diameter (miles)	Average Distance from the sun (miles)	Min/Max Surface Temperature (deg F)	Day	Year (days)	Equatorial Surface Gravity	% Oxygen in atmosphere
Earth	7,928	92,955,821	-129/136	24 hours	365.25	100% of Earth	21%
Mars	4,220	141,633,260	-207/81	24 hours, 40 min	686.98	38% of Earth	0.13%

Source: Review of U.S. Human Spaceflight Plans Com mittee

Figure 3.3.1-1. Comparison of major features of the Earth and Mars

Source: NASA Hubble Space Telescope

Figure 3.3.1-2. Mars – A possible first destination for exploration

The Mars First scenario represents a comprehensive human exploration strategy, and requires a focused technology development program as well as an integrated test plan to reduce risk while gaining confidence and experience with the Mars exploration systems. Exploration of Mars would be performed in extended stays on the surface, with each mission going to a different landing site on Mars. Human exploration would be complemented by robotic exploration of Mars. Synergies would be exploited, but would not fundamentally drive the program.

This scenario was analyzed based on the existing 2007 NASA Human Exploration of Mars Design Reference Architecture 5.0 (NASA-SP-2009-566 and NASA-SP-2009-566-

ADD). This architecture is shown in Figure 3.3.2-1. It assumed the use of eight or more Ares V launchers plus an Ares I crew launch for each Mars opportunity. Both nuclear thermal rockets and chemical (LOX/LH2) in-space propulsion systems were examined. Under this scenario, 26 months before the launch of the crew, a mission is launched carrying the ascent vehicle to the Martian surface, and a second mission injects a crew descent and habitation vehicle into Mars orbit. The crew then makes a faster trip, and reaches rendezvous in Mars orbit with the descent vehicle, lands on the surface of Mars, spends 540 days on the surface, and then returns to Mars orbit in the ascent vehicle. There the crew rendezvouses with the crew vehicle and returns home to Earth. On both the trip to and from Mars, the crew is exposed for 180 to 200 days to the weightlessness and full solar and galactic cosmic radiation of free space (i.e., away from any planet). The total round trip to Mars and back lasts about 900 days.

The Moon would be used as a site for integrated testing of the Mars systems. The systems landed on the Moon—landers, habitat, rovers and other surface systems—would be those designed for the Mars missions, or as similar as possible while suitable for the lunar surface, as shown in Figure 3.3.2-2. Not all systems necessary for Mars can be tested on the Moon (e.g., entry systems and in-situ atmospheric resource utilization). However, most of the Mars systems can be tested on the Moon, making the Moon not only a conceptual testbed, but also an actual testbed for Mars systems.

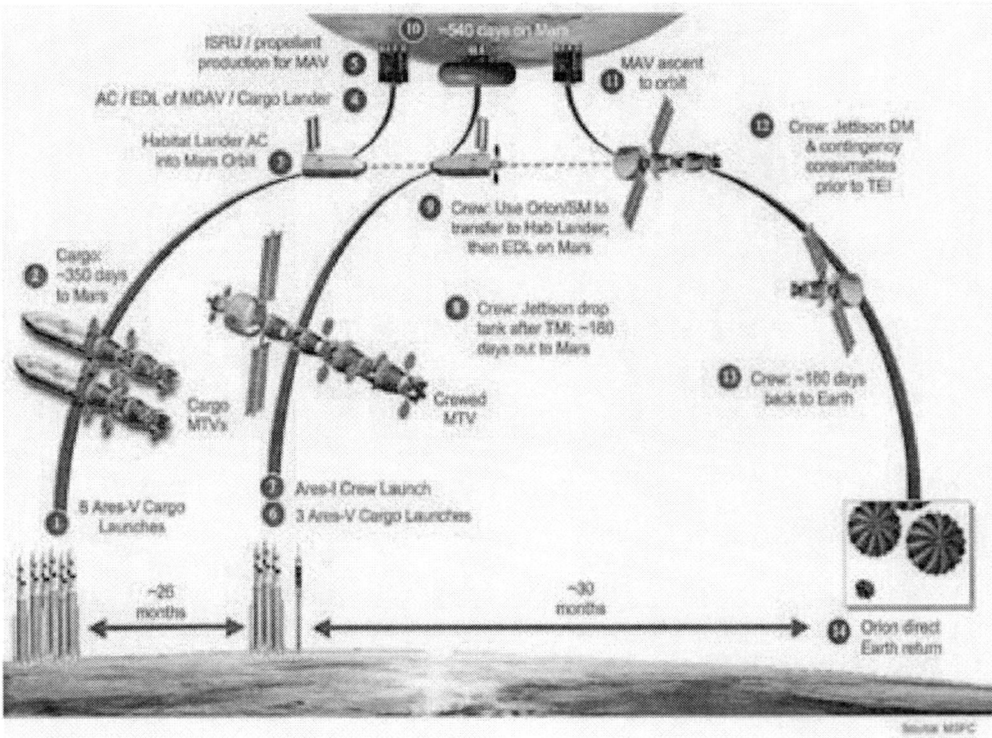

Source: NASA

Figure 3.3.2-1. Architecture of the Mars First strategy, indicating the three missions launched toward Mars necessary to support the landing of a crew of six astronauts

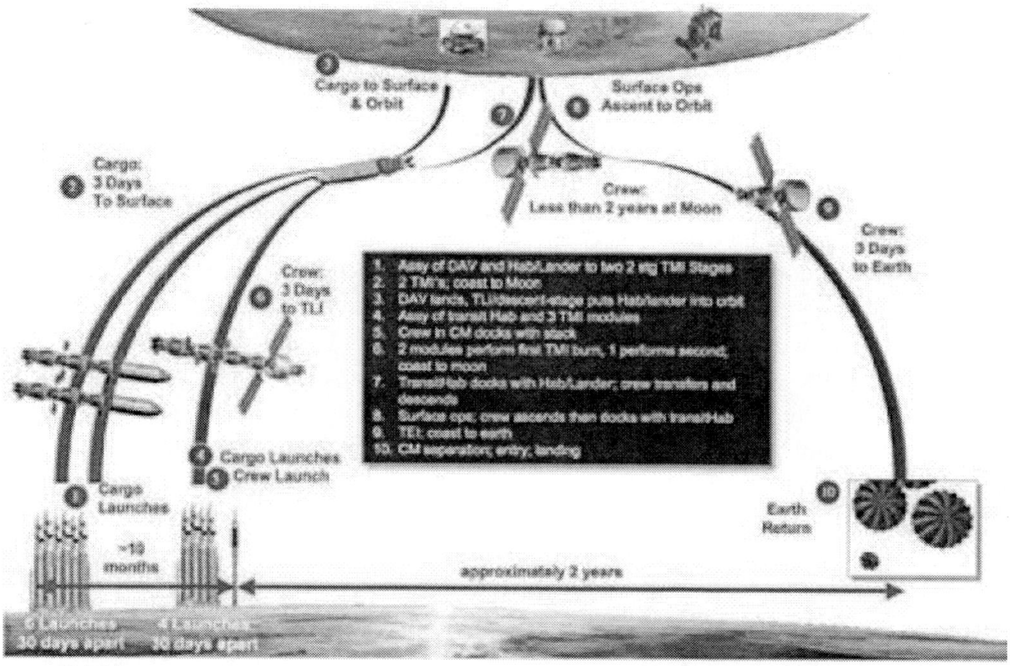

Source: NASA

Figure 3.3.2-2. Architecture of the Mars First system being tested on the Moon

3.3.3. Milestones, destinations and capabilities

A notional development plan and flight road map for the Mars First scenario is presented in Figure 3.3.3-1. The strategy shows the progressive expansion and extension of human capabilities, starting with a demonstration of an extended-dura tion habitat at the ISS, expanding to unpiloted missions to near-Earth objects to demonstrate the performance of heavy lift and in-space transportation systems. At the same time, sub-scale robotic missions at Mars would demonstrate key Mars exploration technologies needed to land large payloads at a precise location. The lunar dress rehearsal would take place as the Mars hardware is completed and prepared for integrated systems testing. The transportation system, cargo and crew would be launched toward the Moon over a 10-month period. Once the crew and cargo are in place on the surface, operations will take place over a two-year period, with crew rotation occurring every six months. While going to the Moon would primarily be for the purpose of testing Mars systems, astronauts would explore the lunar surface as well. When all systems have been tested, cargo-only and then piloted missions to Mars would occur on successive opportunities, about 27 months apart.

3.3.4. Assessment

While Mars is the ultimate destination for the near-term human exploration of space, it is not an easy place to visit with existing technology and experience. No human has ever traveled more than three days from Earth, and none beyond 386 miles away for almost 40 years. No American has been in space much more than 180 days at a time, or exposed to the full radiation of free space for more than about a week. Mars requires a trip in space of almost 900 days. We do not have flight-demonstrated technology to confidently approach and land large spacecraft on the Mars surface. Mars is distant enough from the Sun that it is a weak energy source, and space-based surface nuclear power is probably needed. Under current plans, as many as 12 Ares V vehicles would be needed to launch each biannual set of missions. It seems likely that some form of advanced propulsion may also be needed to make travel feasible. A focused technology program almost a decade long would be required before system design could begin.

The preliminary estimates of the cost of Mars missions are far higher than for other scenarios, all in an era when budgets are becoming highly constrained. If astronauts were to travel to Mars under these circumstances, it would require most of the human spaceflight budget for nearly two decades or more, and produce few intermediate results. When we finally reached Mars, we might be hard pressed to maintain the financial resources needed for repeated missions after the first landings, recreating the pattern of Apollo. For these reasons, the Committee found that Mars is the ultimate destination for human exploration of the inner solar system, but is not a viable first destination beyond low-Earth orbit.

Findings on Human Missions to Mars

Mars as the ultimate destination

Mars is the ultimate destination for human exploration of the inner solar system. It is the planet most similar to Earth, and the one on which permanent extension of human civilization, aided by significant in-situ resources, is most feasible. Its planetary history is close enough to that of the Earth to be of enormous scientific value, and the exploration of Mars could be significantly enhanced by direct participation of human explorers.

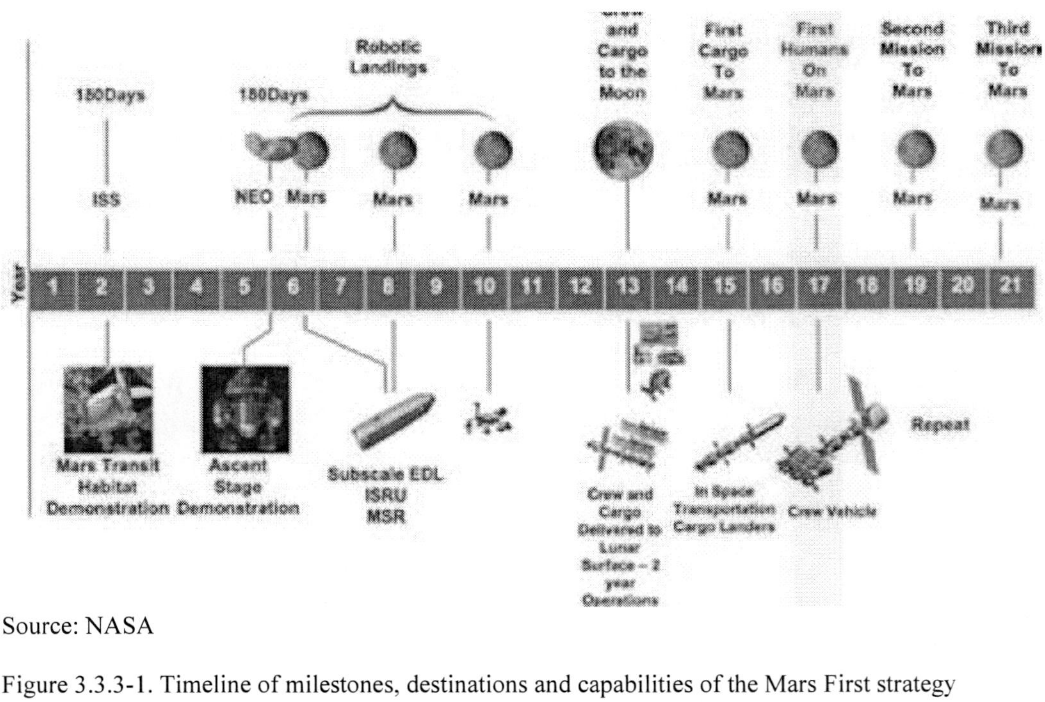

Source: NASA

Figure 3.3.3-1. Timeline of milestones, destinations and capabilities of the Mars First strategy

Mars as the first destination ("Mars first")

Mars is not a viable first destination beyond low-Earth orbit at this time. With existing technology and even a substantially increased budget, the attainment of even symbolic missions would demand decades of investment and carry considerable safety risk to humans. It is important to develop better technology and gain more experience in both free space and surface exploration prior to committing to a specific plan for human exploration of the surface of Mars.

3.4. Moon First

3.4.1. Overview

If Mars is not the first destination be-yond low-Earth orbit, the Moon is an obvious alternative. (See Figure 3.4.1-1.) Going there would enable the development of the operational skills and technology for landing on, launching from and working on a planetary surface, as well as providing a basis for understanding human adaptation to another planet that would one day allow us to go to Mars. Systems would be designed for the Moon, but would be as extensible as practicable for use on Mars. At a minimum, they would demonstrate technologies and operational concepts that would be incorporated into eventual Mars systems.

There are potential resources on the Moon that one day could be launched from the Moon to fuel depots at the Earth–Moon Lagrange points, which could then be used by exploration missions beyond the Earth-Moon system. The scientific exploration of the Moon is not, in and of itself, a rationale for human exploration, but our scientific knowledge of the Moon is incomplete. Our previous missions to the Moon, both human and robotic, encompassed a geographically limited number of sites for a limited time, with little surface range. Much remains to be learned.

Source: NASA

Figure 3.4.1-1. The Earth's Moon-- the initial destination of the Moon First Strategy

3.4.2. Scenario description

In exploring the Moon, there are two main strategies. Both begin with a handful of approximately week-long sorties to various sites to scout, explore regions of different geography, and validate the lunar landing and ascent systems. The *Lunar Base* strategy would begin with the building of a base, probably at the lunar south pole, where the Sun is visible much of the time (as it is at the Earth's south pole in the austral summer). Over many missions, a small colony of habitats would be assembled, much like the base at the South Pole of the Earth, and explorers would begin to live there for up to 180 days. Larger rovers would begin to explore hundreds of miles from the base. Activities would include science exploration and prospecting for resources of hydrogen-rich deposits that could be used as fuel.

Exploring with a few short-duration sorties of just hundreds of miles from the base would actually only allow exploration of a very small fraction of the surface of the Moon in any depth. The alternate strategy for the Moon is to continue a series of increasingly longer sorties to different sites, spending weeks and then months at each one. The primary feature of this *Lunar Global* exploration strategy is that surface operations are flexible and adaptable, so that as discoveries on the lunar surface are made, future missions can be planned that adjust stay times (from 14 to 180 days) and mobility range capabilities in response to those discoveries. There is even the potential that mobile elements could be relocated from one site to another between human visits.

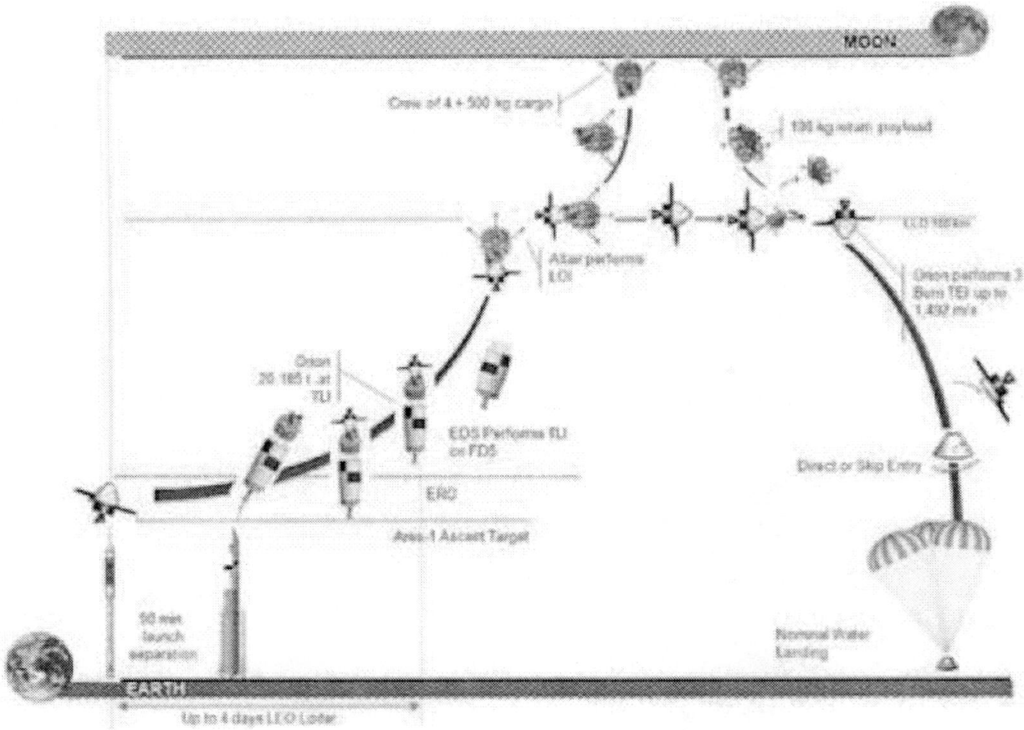

Source: NASA

Figure 3.4.2-1. The architecture of the Moon First scenario, using Ares I and Ares V launchers

The lunar exploration options were informed by the Constellation Program plans. These plans trace to the 2005 Exploration Systems Architecture Study (ESAS) that was established to define an architecture that would comply with guidance from the 2004 Vision for Space Exploration and the 2003 Columbia Accident Investigation Board (CAIB) report. Since ESAS, the human spaceflight program continued technical trades, culminating in the 2008 Lunar Capabilities Concept Review, an architecture-level review that brought together the performance, cost, risk and schedule of the transportation architecture and verified that representative lunar surface mission goals could be accomplished.

The Constellation Program's "1.5 launch" Earth Orbit Rendezvous-Lunar Orbit Rendezvous transportation architecture is shown in Figure 3.4.2-1. The flight elements are launched on two separate vehicles: the Ares I for launch of the Orion spacecraft and crew to low- Earth orbit, and the larger Ares V for launch of the Altair lunar lander and the Earth Departure Stage (EDS). The Orion docks with the Altair/EDS in low- Earth orbit, and the EDS performs the trans-lunar injection burn to send the crew to the Moon. After a four-day coast to the Moon, the Altair descent module engine performs the lunar orbit insertion maneuver, the crew transfers to the Altair, and lands on the surface of the Moon. Following the surface stay, the Altair transports the crew back into orbit for rendezvous with the Orion, which returns to Earth.

The Committee developed the following transportation options to support the lunar scenarios:

- Constellation "1.5 launch" architecture – one Ares I with Orion, plus one Ares V with the Altair lander. This combination is Integrated Option 3 in Chapter 6.
- Ares V Lite "dual" architecture – two Ares V Lites, one with the Orion, and one with the Altair lander. This combination is Integrated Option 4A in Chapter 6.
- A more directly Shuttle-derived launcher, which requires three launches for a crew mission plus one commercial launch of crew to low-Earth orbit. This combination is Integrated Option 4B in Chapter 6.

3.4.3. Milestones, destinations and capabilities

The milestones and destinations of the Lunar Base and Lunar Global alternatives differ slightly. Both begin with a set of sorties to various locations on the lunar surface that enable up to four crew members to explore a single site anywhere on the Moon for up to seven days. This type of mission is accomplished independently of pre-positioned lunar surface infrastructure such as habitats or power systems. The lunar sorties allow for exploration of high-interest science sites, scouting of future lunar outpost locations, or other technology development objectives. The Lunar Sortie mission may include surface mobility assets and science packages, which the crew can operate on daily extra-vehicular activities (spacewalks).

The Lunar Base alternative proceeds then to the construction of a lunar outpost. An example of such a scenario is shown in Figure 3.4.3-1. It develops extensive surface roving capabilities, semi-permanent occupancy by a crew of four, and resource extraction and utilization capability. In later years, it permits the development of additional infrastructure, allowing for science exploration with additional sorties and longer-range roving, and developing operational concepts and experience for Mars exploration.

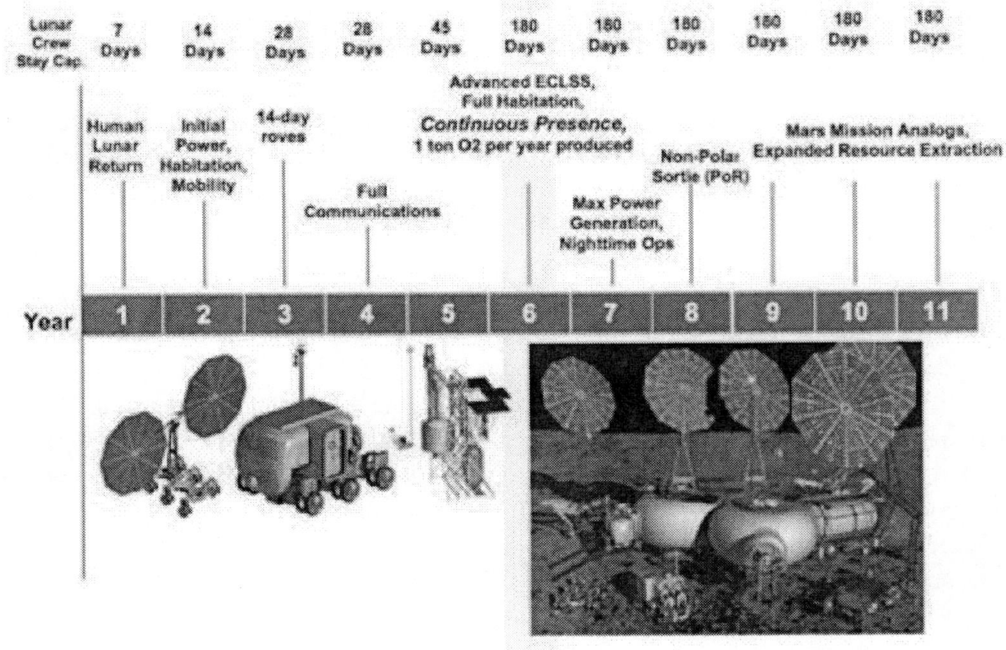

Source: NASA

Figure 3.4.3-1. Timeline of milestones, destinations and capabilities of the Lunar Base variant of the Moon First strategy

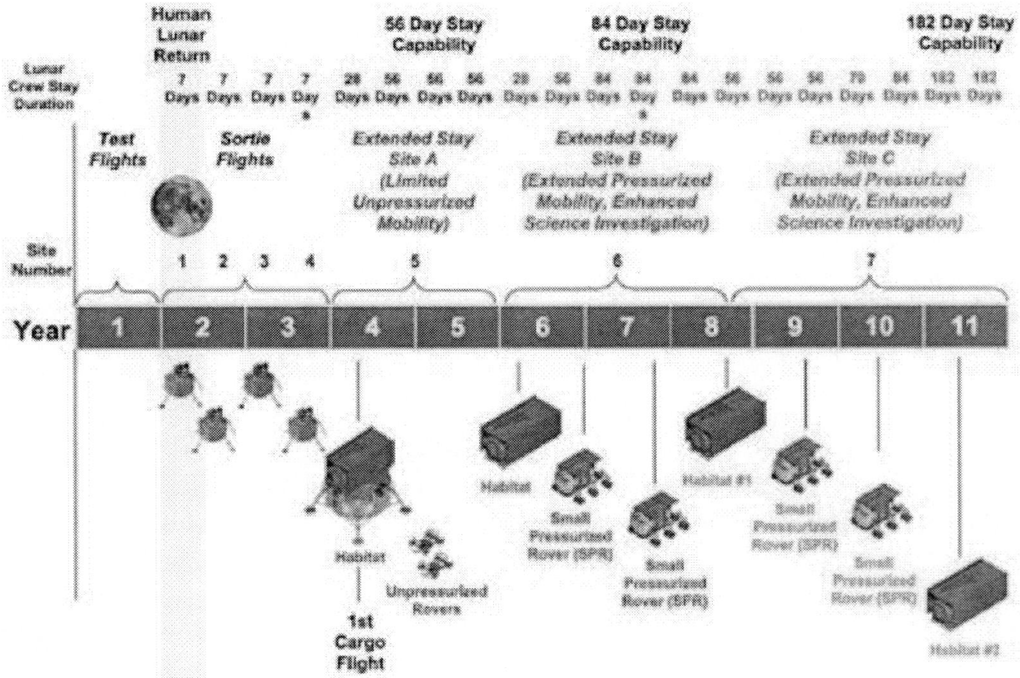

Source: NASA

Figure 3.4.3-2. Timeline of milestones, destinations and capabilities of Lunar Global variant of the Moon First strategy

In contrast, the Lunar Global alternative proceeds from seven-day sorties to longer-duration visits to two to four sites of particular interest, as shown in Figure 3.4.3-2. The first long-duration site would have about a 56 day visit, spending the lunar night on the surface for the first time, and exploring with unpressurized rovers. Subsequent sites would be visited for longer durations, and with more capable exploration infrastructure, they would eventually reach the same 180- day stay as with the Lunar Base. After some number of extended sorties, lunar exploration could be ended, continued in extended sortie mode, or transitioned to a base approach.

3.4.4. Assessment

The Moon is a viable first destination for human exploration beyond low-Earth orbit. Lunar exploration would allow the development of the capability to land on and explore a planetary surface, while still remaining only about three days away from Earth. There are scientific objectives that could be met while visiting the Moon, including studying the evolution of the Moon and using the surface of the Moon as a record for studying the evolution of the solar system. The Moon could compete with other locations as a site for observatories and reduced-gravity surface science. It is worth noting, however, that prior to the announcement of the Vision for Space Exploration in 2004, only one site on the Moon (the south pole Aitken Basin) was on the high- priority locations for robotic exploration of the inner solar system. There is useful science to be performed on the Moon, but science is not the driver of human lunar exploration.

The exploration of the Moon should be focused to the greatest extent possible on developing the technologies and concepts that will be important in further exploration of Mars: surface descent, landing and ascent; habitation and surface exploration; and resource utilization. The Committee explored two strategies for exploring the Moon. The Lunar Base and Lunar Global strategies each have strengths and weaknesses. The Lunar Base allows more efficient utilization of the material brought to the surface, and surface, important to Mars preparation. The Lunar Global approach visits more sites in depth, and more closely simulates the exploration strategy likely to be used on Mars. The Committee finds that the Lunar Base and the Lunar Global exploration strategies have similar costs, and both provide value in exploring the Moon and preparing for the exploration of Mars.

If explored with either of these Moon First strategies, the Moon would help develop some of the technologies and operations concepts needed for Mars exploration, and it would develop some of the transportation infrastructure. However, the Moon does not serve as a perfect analog for Mars. While some of the technology and concepts would be applicable to Mars, most of the major system components (landers, habitats, rovers, etc.), if designed and optimized for the Moon, would have to be redesigned and re-validated for Mars. Other components of the Mars exploration system (atmospheric entry and in-situ resource use, for example) have no analog on the Moon. A long-duration exploration of the Moon is a step towards Mars, but not a giant step, and not the only possible step.

Finding: The Moon as a First Destination ("Moon First")

The Moon is a viable first destination for exploration beyond low-Earth orbit. It initially focuses next steps on entering and departing deep gravity wells, and developing human operations on the surface of a celestial body, which should be developed in a manner that leads to the eventual exploration of Mars. The Moon is nearby, allowing relatively rapid return to Earth in the event of emergencies, and communication transit times are minimal. It also has interesting scientific and resource issues that can be pursued through human exploration.

3.5. The Flexible Path to Mars

3.5.1. Overview

In addition to Mars First and Moon First, there is a third possibility for initial exploration beyond low-Earth orbit: visiting a series of locations and objects in the inner solar system, which the Committee calls the Flexible Path. (See Figure 3.5.1-1.) The goal is to take steps toward Mars, learning to live and work in free space and near planets, under the conditions humans will meet on the way to Mars. We must learn to operate in free space for hundreds of days, beyond the protective radiation belts of the Earth, before we can confidently commit to exploring Mars. Human exploration along the Flexible Path would also support science, create new industrial opportunities, and engage the public through progressively more challenging milestone accomplishments.

On this path, sites would be visited that humans have never reached before. Astronauts would learn to service spacecraft beyond low-Earth orbit, much as crews successfully serviced the Hubble Space Telescope in low-Earth orbit. Humans could visit small bodies in space, such as near- Earth objects (asteroids and spent comets that cross the Earth's path, some of which could someday collide with the Earth - Figure 3.5.1-2) and the scientifically interesting moons of Mars, return samples, and understand their structure and composition. When humans would come close to the Moon or Mars, they could deploy probes and coordinate with or control robotic assets on the surface. They could even bring home samples from Mars that were launched from the surface by robotic spacecraft. In this way we could achieve the scientific "first" of a Mars sample return.

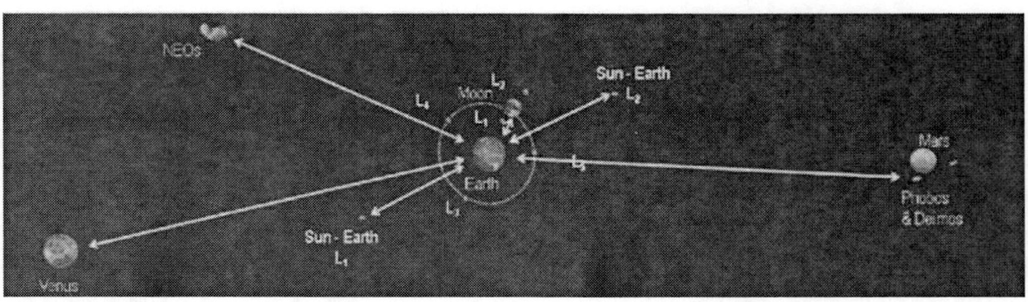

Source: NASA

Figure 3.5.1-1 .There are a variety of destinations that can be targeted using the Flexible Path exploration strategy

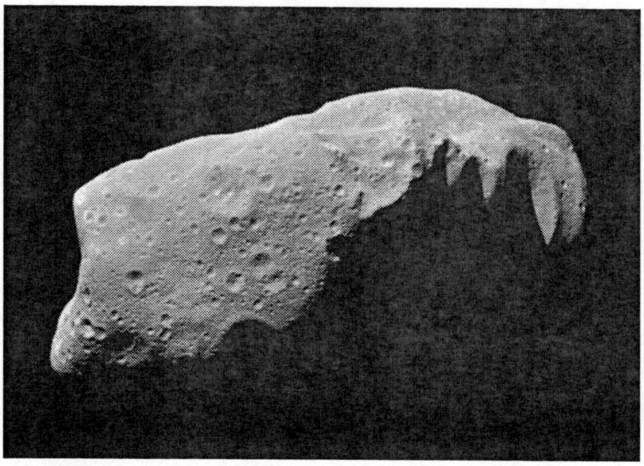

Source: NASA Galileo Satellite

Figure 3.5.1-2. Asteroid Ida – representative of near-Earth objects which are possible destinations along the Flexible Path

These destinations require the smallest energy expenditure beyond low-Earth orbit, but are of increasing distance and duration from Earth. The missions could include a full dress-rehearsal for a Mars mission, consisting of traveling to Mars orbit and returning hundreds of days later. The essential concept is that humans would first visit points in space and rendezvous with small bodies and orbit larger ones, without initially descending into the deep gravity wells of Mars or the Moon.

The Flexible Path is a road toward Mars, with intermediate destinations. At several points along the way, the off-ramp from the Flexible Path to a Moon exploration program could be taken. Alternatively, if new discoveries drew us to Mars, the lunar stop could be bypassed, leading directly to a Mars landing.

3.5.2. Scenario description

The Flexible Path constitutes a steadily advancing, measured, and publicly notable human exploration of space beyond Earth orbit that would build our capability to explore, enable scientific and economic return, and engage the public. The focus of the Flexible Path is to gain ever-increasing operational experience in space, growing in duration from a few weeks to several years in length, and moving from close proximity to the Earth to as far away as Mars.

Humans need to build the capability to explore other planets, and to operate far from the Earth. On the Flexible Path, critical scientific and technological components of human spaceflight would be addressed through incrementally more aggressive exploration missions. Determining the human physiological and operational impacts of (and the counter-measures to) long-term radiation environment (including galactic cosmic rays) and extended exposure to zero-gravity is necessary for a sustained human-exploration capability. The missions would build preparedness to explore by performing increasingly more complex in-space operations and by testing new elements.

The Flexible Path targets planetary scientific return focused on multiple locations in the inner solar system. The goal focuses human exploration on producing exciting new science at each step of the way. The emphasis would be on obtaining multi-kilogram samples from a variety of solar system y bodies through tele-robotic exploration in concert with the human missions. In the case of the Moon and Mars, humans would remain in orbit. They would deploy probes, teleoperate surface robotic vehicles, and potentially rendezvous with sample returns from the surface. In the case of smaller objects, humans would explore the surfaces directly and return samples. Robotic missions would play a visible and complementary role to human exploration through precursor missions and scientific missions that deliver instruments.

A sustained exploration program by the United States requires continuous public engagement, inspiration and benefit. The Flexible Path missions are designed in part to cultivate and maintain public support and interest in human spaceflight by taking on useful, demonstrably new, high-profile missions. They start with unprecedented space missions offering dramatic perspectives of humankind's home planet as a member of the inner solar system, including near-Earth objects, the Moon and Mars.

A set of missions along the Flexible Path might include early visits to lunar orbit, stops at the Earth's Lagrange points, near-Earth objects and visits in the vicinity of Mars. A more detailed sequence might include:

- Orbiting the Moon to learn how to operate robots on a planet from orbit (days in duration).
- Visiting the Earth-Moon and Earth-Sun Lagrange points (special points at the edge of Earth's influence), which are likely sites for science spacecraft servicing and potentially important for interplanetary travel. Earth- Moon Lagrange points are about 85 percent of the way to the Moon from the Earth. Earth-Sun Lagrange points are about four times as far from the Earth as the Moon (weeks to months in mission duration).
- Visiting several near-Earth objects (asteroids or burned-out comets whose path cross the Earth), to return samples and practice operation near a small body and potentially practice in-situ resource extraction (months in mission duration).
- A fly-by of Mars, demonstrating distant operation and coordination with robotic probes on the planetary surface and during rendezvous (years in mission duration).
- A trip to Mars orbit, rendezvousing with and returning samples from Mars's moons (Deimos and Phobos), and potentially from Mars's surface (years in mission duration).

A more detailed explanation of the activities at each destination is shown in Figure 3.5.2-1.

Key assumptions for the Flexible Path scenarios include the notion that viable and relevant exploration missions can be completed with a single crew launch and a single in-space propulsion stage. As additional deep-space capabilities, such as in-space habitats, air locks and propulsion stages, become operational, the scope of the missions increases. Flexible Path missions assume the development of certain enabling technologies. First among these is a cryogenic in-space propulsion stage able to have a near-zero boil-off of propellant over almost 200 days, equipped with a high performance in-space re-startable engine. Additional enabling technologies are in-space cryogenic fluid transfer, improved regenerative life-support systems, technologies for deep space crew-system operational autonomy, and tele-robotic systems to be operated by the crew in deep space.

Destination	Public Engagement	Science	Human Research	Exploration Preparation
Lunar Flyby/Orbit	Return to Moon, "any time we want"	Demo of human robotic operation	10 days beyond radiation belts	Beyond LEO shakedown
Earth Moon L1	"On-ramp to the inter-planetary highway"	Ability to service Earth Sun L2 spacecraft at Earth Moon L1	21 days beyond the belts	Operations at potential fuel depot
Earth Sun L2	First human in "deep space" or "Earth escape"	Ability to service Earth Sun L2 spacecraft at Earth Sun L2	32 days beyond the belts	Potential servicing, test airlock
Earth Sun L1	First human "in the solar wind"	Potential for Earth/Sun science	90 days beyond the belts	Potential servicing, test in-space habitation
NEO's	"Helping protect the planet"	Geophysics, Astrobiology, Sample return	150-220 day, similar to Mars transit	Encounters with small bodies, sample handling, resource utilization
Mars Flyby	First human "to Mars"	Human robotic operations, sample return?	440 days, similar to Mars out and return	Robotic operations, test of planetary cycler concepts
Mars Orbit	Humans "working at Mars and touching bits of Mars"	Mars surface sample return	780 days, full trip to Mars	Joint robotic/human exploration and surface operations, sample testing,
Mars Moons	Humans "landing on another moon"	Mars moons' sample return	780 days, full rehearsal Mars exploration	Joint robotic/human surface and small body exploration

Source: Review of U.S. Human Spaceflight Plans Committee

Figure 3.5.2-1. Benefits of various destinations along the Flexible Path

The Flexible Path branch that proceeds to the lunar surface involves a lander smaller than the Altair lander. For the costed option, it is assumed that NASA would provide the ascent stage, but a commercially acquired descent stage is envisioned that could be developed based on the same in-space re-startable engine discussed above. Several complementary robotic missions are coupled and require some technical development— in particular, the Mars Sample Return mission would use the heavy lift in-space propulsion stage to send multiple sampler missions in a precursor Mars Entry Descent and Landing (EDL) aeroshell to validate technologies for eventual human landings on Mars.

For the missions assessed in this analysis, an Orion capsule was assumed to be capable of carrying up to four crew members and operating in space for over a year. However for missions longer than about a month, an additional in-space habitat sustains the crew. All of the Earth entry, descent, and landings were to fall within the nominal design requirements for the Orion. The three transportation architectures (see Chapter 5) considered for Flexible Path missions are:

- Ares V Lite – an Ares V Lite launches with crew aboard the Orion capsule. This combination is Option 5A in Chapter 6.
- An EELV-heritage super-heavy launcher, which requires two launches for earlier missions, and three launches for later missions. A commercial service transports the crew to orbit, where they transfer to the Orion. This combination is Option 5B in Chapter 6.
- A Shuttle-derived launcher, which also requires two launches for earlier missions, three launches for later missions, and commercial transport of the crew to low- Earth orbit. This combination is Option 5C in Chapter 6.

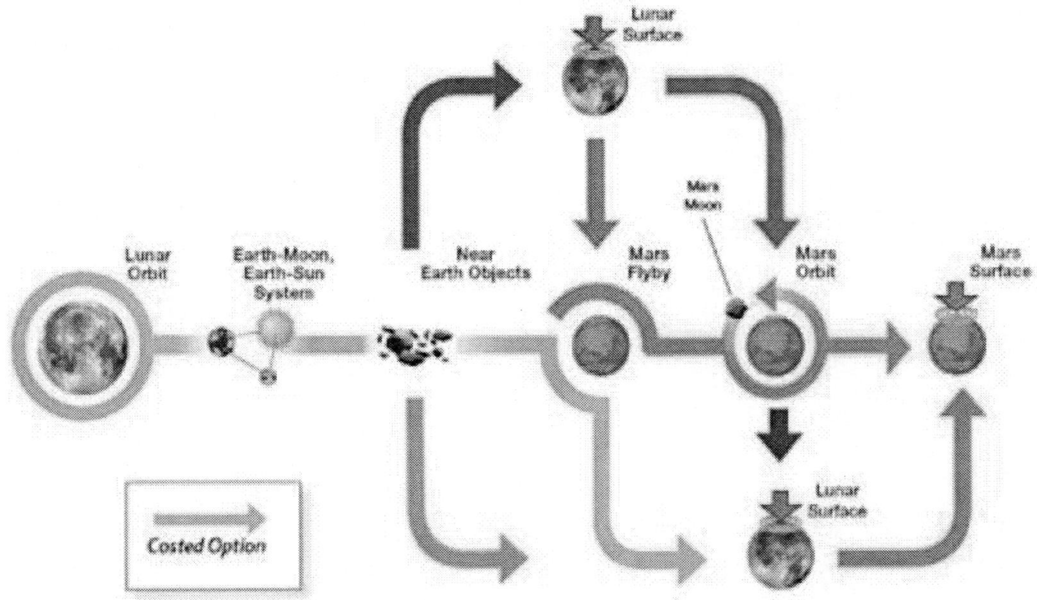

Source: Review of U.S. Human Spaceflight Plans Committee

Figure 3.5.3-1. Options for exploration within Flexible Path strategy showing the main path toward Mars with alternatives to the Moon

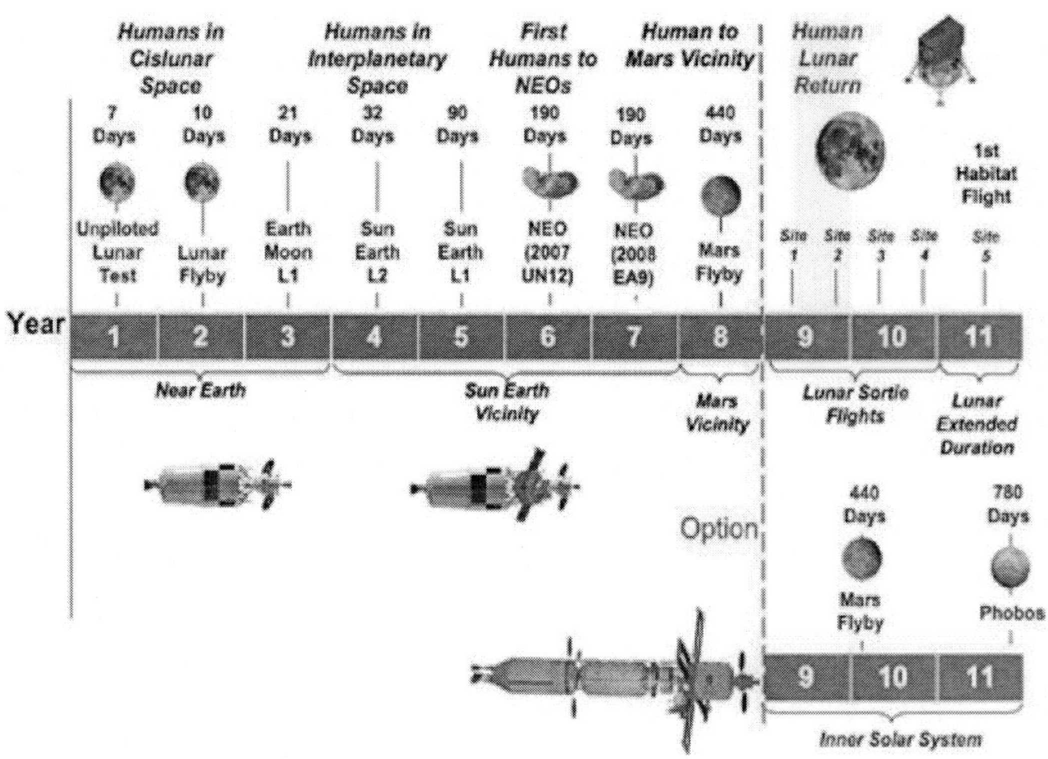

Source: NASA

Figure 3.5.3-2. Timeline of milestones, destinations and capabilities of the Flexible Path strategy

3.5.3. Milestones, destinations and capabilities

The milestones and capabilities for the Flexible Path are best visualized in Figure 3.5.3-1, which shows the flexibility in the strategy. While the Flexible Path missions can be conducted in almost any sequence, many of the missions build upon the experience gained from prior ones. All assume a first flight to lunar orbit, and then to the Lagrange points, and then to near-Earth objects. The various alternatives are then more apparent. The subsequent flow would be to continue into the exploration of Mars, first as a fly-by, then to Mars orbit, and finally a Mars landing. Off-ramps to the Moon could occur at several spots. Mars is the ultimate destination.

A possible sequence that places missions in an order that successively expands capabilities and reduces risk is shown in Figure 3.5.3-2. This sequence follows the path to near- Earth objects, and then performs a Mars fly-by. The main path takes the off-ramp to lunar exploration. This alternative with the lunar off-ramp appears in Chapter 6 among the costed options. Another alternative continues Mars exploration with a mission to Mars's moons. At this point the next obvious step is to develop the systems and technologies to land humans on Mars.

3.5.4. Assessment

The Flexible Path is a viable strategy for the first human exploration of space beyond low-Earth orbit. Humans could learn how to live and work in space, gaining confidence and experience traveling progressively farther from the Earth on longer voyages. This would prepare for future exploration of Mars by allowing us to understand the long-term physical and emotional stress of human travel far from the Earth. It would also validate in-space propulsion and habitat concepts that would be used in going to Mars.

The missions would go to places humans have never been to, escaping from the Earth/Moon system, visiting near-Earth objects, flying by Mars, thereby continuously engaging public interest. Explorers would initially avoid traveling to the bottom of the relatively deep gravity wells of the surface of the Moon and Mars, but would learn to work with robotic probes on the planetary surface. This would allow us to develop new capabilities and technologies for exploring space, but ones that have Earth-focused applications as well. It would also allow us to defer the costs of more expensive landing and surface systems. From the perspective of science, it would demonstrate the ability to service observatories in space beyond low-Earth orbit, as well as return samples from near-Earth objects and (potentially) from Mars.

This flexibility would enable us to choose different destinations, or to proceed with the exploration of the surface of the Moon or Mars. This allows us to react to discoveries that robots or explorers make (such as indications of life on Mars) or eventualities that are thrust upon us (such as a threat from a near-Earth object). It would lead to a better understanding of near-Earth objects, through evaluations of their utility as sites for mining of in- situ resources, and analyses of their structure, should we ever need to deflect one away from the Earth.

Exploration along the Flexible Path would not likely complete our preparation for the exploration of Mars. At some point we would likely need to gain more experience landing and working on an extra-terrestrial planetary surface. This could be done on the Moon with specialized lunar systems, or with systems designed for Mars (as discussed above in the Mars

First option). Alternatively, we could practice autonomous landings of large systems on Mars, in coordination with science programs.

Finding: Destinations on the Flexible Path to Mars

The destinations on the Flexible Path (lunar orbit, Lagrange points, near-Earth objects, Mars fly-bys, Mars orbits, and Mars moons, with potential operation of robotic missions on the lunar and Martian surfaces) comprise a viable and interesting first set of destinations for exploration beyond low-Earth orbit. They are progressively more distant, focusing our next steps on allowing the development and understanding of human operations in free space for the increasingly longer durations necessary to explore Mars. Important scientific, space operational, and Earth-protection benefits would be obtained on this path.

3.6. Summary of Strategies for Exploration beyond Low-Earth Orbit

This chapter addresses the question: What is the most practicable strategy for exploration beyond low-Earth orbit: Mars First, Moon First, or the Flexible Path to Mars?

The Committee found that, although Mars is the ultimate destination for human exploration in the inner solar system, it is not a viable first destination. We do not now have the technology or experience to explore Mars safely and sustainably.

Both the Moon First and Flexible Path are viable strategies. Exploring the Moon would prepare us for exploration of Mars by allowing us to learn to live and work on a remote surface, yet one that is only three days from Earth.

The Flexible Path would prepare us for exploration of Mars by developing confidence that we can live and work in free space, and allowing us to learn to explore planets and bodies in a new way, potentially in coordination with robotic probes.

The Moon First and Flexible Path destinations are not mutually exclusive; before traveling to Mars, we will probably both extend our presence in free space and work on the lunar surface. For example, if we had had explorers on the Moon for a decade, but never more than three days from Earth, would we easily commit to a mission that took our astronauts away for three *years*? This seems unlikely. Likewise, if we had worked in space for a decade, would we commit to landing on a planet 180 days away without practice? This seems equally unlikely.

Before we explore Mars, we will likely do some of both the Flexible Path and lunar exploration—the primary decision *is one of sequence*. This will be largely guided by budgetary, programmatic, and program sustainability considerations. These will be discussed in Chaper 6. Whatever destination(s) are selected, it would be desirable to have a spectrum of program choices that offers periodic milestone accomplishments visible to and appreciated by the public.

Before leaving the topic of strategies for exploration, it is important to reflect back on the goals for human spaceflight. The strategies have certainly focused on preparing for a venture to Mars, and therefore the potential expansion of human civilization into the solar system. Many opportunities could be identified in either pathway for deep involvement of international partners, as will be suggested in Chapter 8. Science returns, technology and economic development, and exploration preparation will follow from either strategy.

The primary question is: Will the public be engaged? Gone is the era when Americans remembered the names of astronauts, or even the date of the next space launch. We cannot take for granted that space excites young people; rather, we must make sure that we build a program that will excite them. In its plan for exploration, NASA must find new ways to engage the public, particularly young people, in a venture of participatory exploration, one that will be exciting to them. This should not be an afterthought—it must be an integral part of the program.

Finding: Pathways to Mars

Mars is the ultimate destination for human exploration of the inner solar system; but it is not the best *first* destination. Both visiting the Moon First and following the Flexible Path are viable exploration strategies. The two are not necessarily mutually exclusive; before traveling to Mars, we might be well served to both extend our presence in free space and gain experience working on the lunar surface.

4. Current Human Spaceflight Programs

The current U.S. human spaceflight programs are the operational Space Shuttle Program and the U.S. portion of the International Space Station (ISS). The next human spaceflight effort, the Constellation Program, is in development.

4.1. The Space Shuttle

The Committee has addressed five questions that, if answered, would form the basis of a plan for U.S. human spaceflight. First among those questions is: what should be the future of the Space Shuttle?

The current plan is to retire the Shuttle at the end of FY 2010. Six flights are remaining on the manifest, with the final flight scheduled for September 2010. Once the Shuttle is retired, there will be a gap in America's capability to independently launch people into space. That gap will extend until the next U.S. human-rated launch system becomes available.

In analyzing the future of the Shuttle, the Committee considered whether the current flight schedule is realistic. It also weighed the risks and possible benefits of various Shuttle extension options. This section provides a brief background on the Space Shuttle, a discussion of important issues, and a description of the scenarios considered for inclusion in the integrated options presented in this report.

4.1.1. Background

The Space Shuttle, introduced in 1981, is fundamentally different from all previous U.S. launch systems. (See Figure 4.1.1-1.) It lifts astronauts to orbit in a spaceplane, not a capsule, and it lands on a runway, not with a splash in the ocean. The spaceplane has a cargo bay to carry satellites and experiments with it into space and back to Earth, and it can be flown again and again.

The Shuttle has been the workhorse of the U.S. human spaceflight program since its first launch. In its 28 years of operations, it has flown 128 times, 126 of those successfully. Two tragic accidents mar its record. Space Shuttle missions have evolved considerably in focus, capability and complexity over that period. They have progressed from early flight tests to operations, which included satellite deployments, tests of a robotic arm, and early scientific experiments. Immediately after the Space Shuttle *Challenger* accident in 1986, the launch of satellites shifted from the Shuttle to expendable launch vehicles, and Shuttle missions evolved into more sophisticated science and operational missions, including Spacelab flights, repair and servicing of the Hubble Space Telescope, and the Shuttle Radar Topography Mission.

In the late 1990s, the focus of Shuttle missions transitioned to the assembly, logistics support, and maintenance of the International Space Station. The Space Shuttle *Columbia* accident in early 2003 interrupted that work, grounding the Shuttle for nearly two and a half years while NASA addressed the technical, procedural and organizational problems identified during the accident investigation. When the Shuttle returned to flight, its missions concentrated almost entirely on completing assembly of the Space Station. The President's 2004 Vision for Space Exploration directed NASA to: "Focus use of the Space Shuttle to complete assembly of the International Space Station; and retire the Space Shuttle as soon as assembly of the International Space Station is completed, planned for the end of this decade."

Source: NASA

Figure 4.1.1-1. The principal components of the Space Shuttle

Subsequently, several Shuttle flights planned to support the International Space Station assembly and utilization were cancelled, and NASA was directed to complete the remaining Shuttle flights by the end of FY 2010. At the time, the Constellation Program's replacement for the Shuttle was projected to be ready in 2012, leaving a two-year "gap" in the nation's ability to launch humans into low-Earth orbit.

As of the end of FY 2009, the Shuttle has flown successfully 15 times since returning to flight in 2005. Missions are now far more intricate and complex than earlier Shuttle flights, and they illustrate significant growth in the ability to operate in space. While early missions were routinely four to seven days, and rarely included a spacewalk, missions today are often two weeks long, and have included as many as five complex and well-orchestrated spacewalks. As of September 2009, six flights remain in the Shuttle manifest, with the last flight scheduled for September 2010. There is currently modest funding in the FY 2011 budget to cover Shuttle retirement costs, but none for flight operations.

4.1.2. Issues

In considering the future of the Space Shuttle, the Committee paid particular attention to safety, schedule, workforce, and the program's fixed costs.

Schedule To assess the viability of the current Shuttle schedule, the Committee compared the actual post-*Columbia* flight rate (July 2005 through STS-128, the last flight in FY 2009) with the projected flight rate for the remainder of the current manifest. In the post-*Columbia* period through the end of FY 2009, there was an average of 100 days between flights. In contrast, the current manifest shows an average of only 64 days between the remaining six flights. While it is not impossible to achieve this latter flight rate, the projected rate is not consistent with recent or prior experience. Further, Space Shuttle managers have indicated that there is little or no margin in the remaining schedule. Experience suggests that it is very likely the currently manifested flights will extend into the second quarter of FY 2011.

The Committee also took note that the Columbia Accident Investigation Board (CAIB) cited schedule and budget pressure as a contributing factor in the *Columbia* accident. The Board observed, "Little by little, NASA was accepting more and more risk in order to stay on schedule." It recommended that NASA: "Adopt and maintain a Shuttle flight schedule that is consistent with available resources" and added that "Although schedule deadlines are an important management tool, those deadlines must be regularly evaluated to ensure that any additional risk incurred to meet the schedule is recognized, understood, and acceptable."

NASA and the Committee are well aware that schedule pressure can have a subconscious influence on decision- making, and has asked for, and received, relief from the requirement to fly out the manifest by the end of FY 2010. The Administration has directed that NASA complete the remaining manifest safely, even if that requires extending into FY 2011. The looming problem, however, is that there is currently no funding in the FY 2011 budget to support this likely occurrence.

Workforce The most visible ramification of the impending gap in U.S. human spaceflight is the lengthy loss of ability for the U.S. to launch humans into space independently. A less well-publicized ramification is the potential loss of the knowledge and skill base that makes America's human spaceflight program possible.

The Space Shuttle is currently operated by a skilled workforce of over 12,500 individuals whose experience and expertise in systems engineering, systems integration, inspection, ground operations and assembly, test and checkout, and mission planning and operations have been developed and honed over decades. Once the Shuttle is retired, NASA and its contractors will be forced to shed or reassign much of that workforce due to the length of the gap in human spaceflight activity. Of these 12,500 workers, 1,500 are civil servants who, under current practices, will likely retain their jobs even though there is no program to which they can easily transition. The jobs in the contractor structure will likely be lost. (See Figure 4.1.2-1.) When the human spaceflight program resumes in the second half of the next decade, a great deal of the knowledge, experience and critical skills necessary for successful program execution is likely to have atrophied or have been lost altogether. Over the past 45 years, the U.S. has enjoyed a relatively continuous program of human spaceflight. This continuity enabled engineers, flight operations personnel and technicians to learn skills and train successors in an apprentice model, and to capture and transfer knowledge from one program to the next. The longest previous gap occurred between the Apollo-Soyuz mission in 1975 and the first Shuttle flight in 1981. But even as late as 1977, the Shuttle was projected to fly in 1979. As a result, only a year or two after Apollo-Soyuz flew, much of the workforce was actively engaged in ground processing, systems engineering, integrated testing, flight crew training and mission planning for Shuttle.

The Committee is concerned about the retention of critical knowledge and skills and the availability of that unique portion of the workforce necessary to conduct the next set of human spaceflight missions—which, as of now, cannot be expected until late in the next decade.

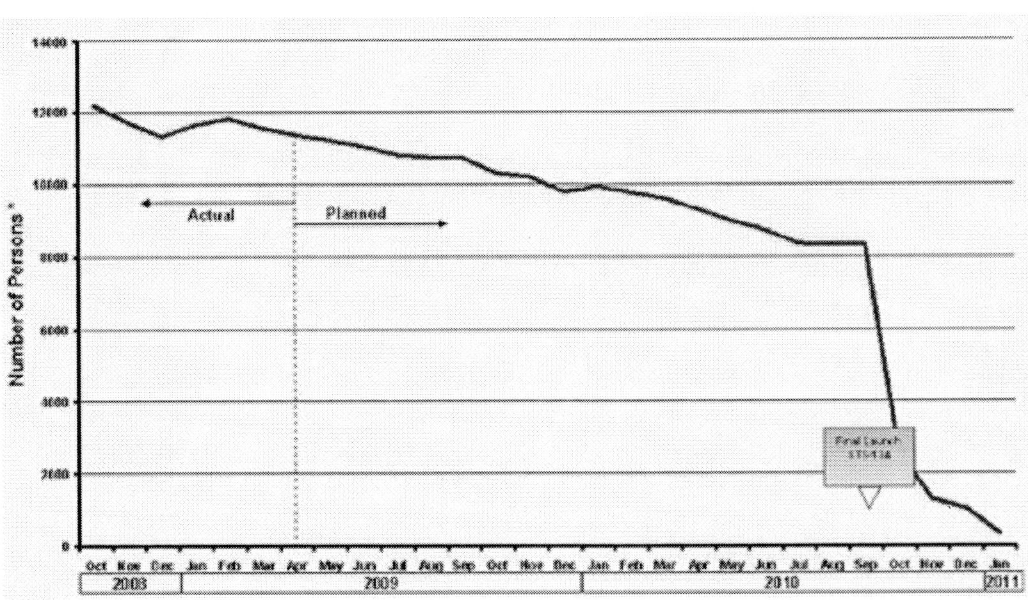

Note: *Includes the workforce for the four largest Space Shuttle contractors only (not NASA space employees).

Figure 4.1.2-1. Space Shuttle Program contractor workforce under existing plan. Source: Briefing to the Review of U.S. Human Spaceflight Plans Committee

Safety The Committee's charter did not call for it to review the safety record or assess the reliability of the Shuttle. The Committee did, however, consider Shuttle safety and reliability in its deliberations. One of the recommendations of the Columbia Accident Investigation Board (CAIB) spoke directly to this issue: "Prior to operating the Shuttle beyond 2010, develop and conduct a vehicle recertification at the material, component, subsystem, and system levels. Recertification requirements should be included in the Service Life Extension Program."

As part of the Shuttle "Return to Flight" program after the Columbia accident and in the years since, NASA has recertified much of the Shuttle system. NASA's Space Shuttle program managers believe the program is meeting the intent of the CAIB 's recommendation and would be ready to fly the Shuttle beyond 2010, should the need arise. This Committee suggests that an independent review of the Shuttle recertification process be undertaken if a decision is made to add flights to the current manifest.

How reliable and how safe is the Shuttle, particularly when compared to other existing or proposed launch vehicles? As noted previously in this report, flying in space is inherently risky, so it is not appropriate to call any launch vehicle "safe." Several factors contribute to a launch vehicle's risk: the design itself; the extent to which the limitations of that design are understood; the processes and people involved in preparing, launching and operating the vehicle; and "random" component or system failures. Studies of risk associated with different launch vehicles (both human-rated and non-human-rated) reveal that many accidents are a result of poor processes, process lapses, human error, or design flaws. Very few result from so-called random component failures. The often-used Probabilistic Risk Assessment (PRA) is a measure of a launch vehicle's susceptibility to these component or system failures. It provides a useful way to compare the relative risks of mature launch vehicles (in which the design is well understood and processes are in place); it is not as useful a guide as to whether a new launch vehicle will fail during operations, especially during its early flights.

The Shuttle is one of the few launch vehicles that have flown a sufficient number of times to be considered "mature." It has suffered two accidents in its 128 flights, so its demonstrated success rate is 98.4 percent. Considerable effort has also been expended to develop a Probabilistic Risk Assessment for the Shuttle. That PRA shows a reliability of 98.7 percent, with the greatest contributor to risk coming from the threat of micrometeorite or debris damage while in orbit. Other launch vehicles in development have better PRAs, indicating that once they reach maturity, they will carry less risk than the Shuttle. In comparing Shuttle reliability to that of other launch vehicles, however, the most important factor is actual flight experience. The Shuttle completed its first 24 missions successfully before the *Challenger* accident; after returning to flight, it flew successfully 87 times before the *Columbia* accident, and has flown successfully 15 times since. This is not to say that future vehicles will not be more reliable—they likely will be—but the Shuttle has reached a level of maturity that those launch vehicles will not reach for many years. (Those vehicles still have their "infant mortality" phase ahead of them. The Committee cannot resist citing one of Augustine's Laws: "Never fly on an airplane with a tail number less than 10!" That law encapsulates the value of flight experience.[1])

The current program ensures that we will have no failures of U.S. government human-rated crewed launch systems from 2011 through at least 2017—because there likely will be no flights of those launch vehicles during that period. The Committee considered whether the risk associated with extending Shuttle operations is appropriate. In doing so, it considered whether it is acceptable to complete the current manifest and, if so, whether the risk is acceptable for some number of additional flights (assuming the current level of attention to mission assurance, processes and procedures is maintained.) The Committee believes the risk of flying out the current Shuttle manifest is consistent with past experience if conducted on a schedule and budget that do not impose undue pressure and constraints. The Committee also believes the risk of some extension beyond the current manifest may be acceptable, assuming the certification process discussed above is successfully completed and the current emphasis on mission assurance is continued.

Fixed Costs The annual Shuttle budget is approximately $3 billion per year, depending on the number of flights. The retirement of the Shuttle is expected to free funds for the Constellation Program, and the common perception is that with the Shuttle no longer flying, there will be an additional $3 billion per year available for design, development, testing and deployment of the new exploration program. The situation is more complicated, however, and the actual benefit to the Constellation Program is considerably less than $3 billion per year. The principal reason is that the Shuttle Program today carries much of the costs of the facilities and infrastructure associated with the human spaceflight program as a whole. But those facilities will continue to exist after the Shuttle is retired—so their costs must still be absorbed if the facilities are to be preserved.

These fixed costs are significant—about $1.5 billion per year—and include, for example, nearly 90 percent of the costs of running: the Kennedy Space Center; the engine test facilities at the Stennis Space Center in Mississippi; a Mission Control Center in Houston; and the Michoud Assembly Facility in Louisiana. Unless such facilities are mothballed or disposed of, these costs will simply transfer to a different NASA program; in fact, most will have to be absorbed by the Constellation Program. During its fact-finding phase, the Committee discovered that approximately $400 million per year of these fixed costs are not yet reflected in the Constellation budget after Shuttle retirement. But the costs do have to be allocated somewhere in the NASA budget, and will certainly affect the overall funding available for exploration. Some of the Shuttle funding pays for NASA civil servants who, absent major layoffs, will simply transition to other spaceflight programs. Constellation will thus gain both human resources and the costs associated with them; in the case of facilities, Constellation will soon be paying for their maintenance.

In summary, the savings resulting from Shuttle retirement are not as great as they may appear. Conversely, the marginal costs of flying the Shuttle are less than implied by the existing bookkeeping. The next human spaceflight program will assume most of the fixed costs; the net funds available for Constellation design, development, test and evaluation (DDT&E) or facilities conversion as a result of Shuttle retirement total about $1.6 billion per year—absent structural changes to NASA.

4.1.3. Shuttle options

The Committee selected three possible Shuttle scenarios to consider for inclusion in the integrated options presented later in this report: flying out the Shuttle manifest (at a prudent rate); adding one flight to provide short-term support for the ISS; and closing the gap by extending Shuttle to 2015 at a minimum flight rate.

- **Scenario 1: Prudent Shuttle Fly-Out.** As noted, the current Shuttle schedule has little or no margin remaining. Scenario 1 is a likely reflection of reality. It restores margin to the schedule, at a flight rate in line with recent experience, and allocates funds in FY 2011 to support Shuttle operations into that fiscal year. Based on historical data, the Committee believes it is likely that the remaining six flights on the manifest will stretch into the second quarter of 2011, and it is prudent to plan for that occurrence and explicitly include the associated costs in the FY 2011 budget.

- **Scenario 2: Short-Term Support for the ISS.** Space Shuttle retirement will have an impact on the ISS (described more fully in a subsequent section). Scenario 2 would add one additional Shuttle flight to provide some additional support for the ISS and ease the transition to commercial and international cargo flights. It could enhance early utilization of the ISS, offer an opportunity for providing more spare parts, and enable scientific experiments to be brought back to Earth. This additional Shuttle flight would not replace any of the planned international or commercial resupply flights.

One obvious question is: "Why add just one flight?" Due to the planned retirement, the Shuttle's external tank production line has been closed recently, and it is not cost- effective to re-open it for a small number of new tanks. However, there is one spare external tank remaining in inventory. This scenario thus envisions using that tank and conducting one additional Shuttle flight in late FY 2011 or early FY 2012.

This scenario requires that funds be put in the in FY 2011 and possibly FY 2012 budget for the one additional Shuttle flight. This minimal extension does not mitigate the workforce transition issues; it does extend U.S. human spaceflight capability, but only by a few months; and it does offer some additional short-term logistical support to the ISS.

- **Scenario 3: Extend Shuttle to 2015 at Minimum Flight Rate.** This scenario would extend the Shuttle at a minimum safe flight rate (nominally two flights per year) into FY 2015. Once the Shuttle is retired, the U.S. itself will no longer have the ability to launch astronauts into space, and will have to rely on the Russian Soyuz vehicle. That gap will persist until a new vehicle becomes available to transport crew to low-Earth orbit. Under the current program, the resulting gap is expected to be seven years or more. This scenario, if combined with a new crew launch capability that will be available by the middle of the 2010s, significantly reduces that gap, and retains U.S. ability to deliver astronauts to the ISS.
- The impending gap also directly affects the ISS, which was designed and built assuming that the Shuttle was available to carry cargo and crew to it and to bring cargo and crew back to Earth. During the gap, the U.S. will pay for U.S. and international-partner astronauts to be carried to and from the ISS by the Russian Soyuz. Cargo, including supplies, spares, experiments and other hardware, will be carried to the ISS by a complement of international and U.S. commercial cargo vehicles. None of these can carry nearly as much as the Shuttle, and only one is projected to be able to bring anything back to Earth. This could limit the full utilization of the ISS. Further, only two of these vehicles have flown—each one only once. Delays could place ISS utilization further at risk, particularly in the early part

of the coming decade. This scenario does not envision replacing any of the planned international or commercial *cargo* launches with Shuttle flights, but rather, enhancing U.S. and international partner capability to robustly utilize the ISS. All commercial and international cargo flights would remain as planned.

The Committee has concluded that the only way to eliminate or significantly reduce the gap in human spaceflight launch capability is by extending the Shuttle Program. If that is an important consideration, then this scenario is one to examine. The scenario also minimizes workforce transition problems, and it retains the skills that currently enable the U.S. to enjoy a robust human spaceflight program. Because this scenario extends the Shuttle's life well beyond 2010, if adopted it should include a thorough review of NASA's safety certification program by an independent committee to ensure that NASA has met the intent behind recommendation R9.2-1 of the Columbia Accident Investigation Board.

Scenario 3 would require additional funding for Shuttle extension. Assuming that many of the current fixed costs must be carried somewhere in the NASA budget, the relevant cost of this option is the marginal cost of flying the Shuttle. There are two factors to consider in estimating this cost. First, if the Shuttle extension is coupled with a strategy to develop a more directly Shuttle-derived heavy-lift vehicle, as opposed to the Ares family, there would be synergy that takes maximum advantage of existing infrastructure, design and production capabilities. Second, since the Shuttle would be available to carry crew to and from the ISS, there would be some savings because the U.S. would not need to purchase Russian Soyuz flights (the present plan).

Most of the integrated options presented in Chapter 6 would retire the Shuttle after a prudent fly-out of the current manifest, indicating that the Committee found the interim reliance on international crew services acceptable. However, one option does provide for an extension of the Shuttle at a minimum safe flight rate to preserve U.S. capability to launch astronauts into space. As Chapter 5 will show, the Committee finds that in the long run, it is important for the U.S. to maintain independent crew access to low-Earth orbit.

Findings Regarding the Space Shuttle

Short-term Space Shuttle planning

The remaining Shuttle manifest should be flown in a safe and prudent manner. This manifest will likely extend operations into the second quarter of FY 2011. It is important to budget for this likelihood.

The human spaceflight gap

Under current conditions, the gap in U.S. ability to launch astronauts into space is most likely to stretch to at least seven years. The Committee did not identify any credible approach employing new capabilities that could shorten the gap to less than six years. The only way to close the gap significantly is to extend the life of the Shuttle Program.

Shuttle extension provisions

If the Shuttle life is extended beyond 2011, an independent committee should assess NASA's Shuttle recertification to assure compliance with the Columbia Accident Investigation Board Recommendation R9.2-1. The investment necessary to extend the Shuttle makes the most sense in the context of adopting a Shuttle- derived heavy-lift capability in place of the Ares family and extending the life of the ISS.

Fixed costs

Because a substantial fraction of the costs of the human spaceflight infrastructure is currently allocated to the Shuttle Program, the savings resulting from Shuttle retirement are not as great as they may appear. If current operating constraints on NASA are maintained, these costs will simply be transferred to whatever becomes the continuing exploration program.

4.2. The International Space Station

The second question the Committee addressed to form the basis of a plan for U.S. human spaceflight was: What should be the future of the International Space Station?

NASA's current plan is to decommission the International Space Station at the end of FY 2015. The Committee believes there is no reasonable path to continue operation of the ISS once U.S. participation ends; thus, de-orbiting the facility in early 2016 will be required for ground safety reasons.

In deliberating the ISS's future, the Committee considered the realism of the current plan, the advantages and disadvantages of that plan, and the advantages and disadvantages of various scenarios that would extend the life of the ISS. This section provides a brief background on the ISS, a discussion of important issues, and a description of the scenarios considered for inclusion in the integrated options presented later in the report.

4.2.1. Background

President Ronald Reagan called for the construction of Space Station *Freedom* (Figure 4.2.1-1) in 1984 as an expression of America's continuing leadership in human spaceflight. With the end of the Cold War, however, the U.S. approach to building the Space Station changed. Space Station *Freedom* became the International Space Station in 1993, when President Clinton encouraged the partnership to invite Russia to join the international group building the Station. (Reference Figure 4.2.2-2) The ISS is among the more complex technological endeavors ever undertaken (some would argue the "most"), involving five space agencies representing 16 nations. Soon to be completed, this new outpost will include contributions from the United States, Canada, Japan, Russia, Brazil, Belgium, Denmark, France, Germany, Italy, the Netherlands, Norway, Spain, Sweden, Switzerland and the United Kingdom. Within the U.S., the ISS effort involves more than 100,000 people in 37 states, including a presence at some 500 contractor facilities.

Source: The Ronald Reagan Presidential Foundation and Library

Figure 4.2.1-1. President Reagan and Prime Minister Thatcher, with model of Space Station Freedom, which later evolved into the International Space Station.

Reorienting the program to facilitate Russian participation was considered a major signal of America's willingness to work with a former adversary. The agreement called for the Space Shuttle and Russian Soyuz to fly crew to the Station, and for the Shuttle and the Russian Progress to resupply the Station. In 1998, Russia's Zarya module was the first to be deployed, and the ISS has been continuously inhabited since 2000. As many as 13 people have occupied the Station and the docked Shuttles at one time. Russian launches sustained the ISS after the *Columbia* accident in 2003 until the Shuttle returned to flight in 2005. Now, in 2009, after nearly 10 years of continuous human habitation with a reduced crew, the ISS supports its full six-person crew. Six more Shuttle missions remain until the ISS construction is completed.

Aside from the Space Station itself, perhaps the most valuable outcome of the ISS Program is the development of strong and tested working relationships among the ISS partners. The partnership resolved numerous technical challenges, withstood changes in governments, policies and budgets, and it survived the *Columbia* tragedy. The imminent completion of the ISS demonstrates that many nations can learn to work together toward a difficult com-mon goal. The effort also expresses a U.S. leadership style adapted to the multi-polar world that emerged after the Cold War.

ISS completion also marks a transition for the conduct of NASA's human spaceflight program, not only because the ISS partners will turn from building the Station to using it, but also because the Space Shuttle is nearing the end of its planned operational life. How will the Station be staffed in the gap between Shuttle decommissioning and the availability of new U.S. launch vehicles? Has NASA made the best arrangements for full utilization of the ISS? For these reasons alone, it is time to reexamine how the United States will use the ISS.

There are other considerations as well. The U.S. made a significant sacrifice in order to complete the ISS and fulfill its obligations to its partners: the science and engineering development program that might have been conducted on the station was curtailed. Perhaps the absence of a significant community of U.S. science users made it easier for NASA to propose discontinuing station operations in 2015. But is it wise to cease operations after only five years of full utilization when the station has been 25 years in planning and assembly?

Would extension of ISS operations from five to at least ten years enable more new ideas, based on today's science and technology, to be introduced through flight on the ISS? When the ISS was first designed, there was little thought about using it to prepare for exploration beyond low-Earth orbit. Can ISS utilization advance exploration goals beyond low-Earth orbit?

4.2.2. Issues

In considering the future of the ISS, the Committee examined issues related to the U.S. human spaceflight gap, cargo and crew resupply and the commercial launch industry, end of ISS life, ISS safety, and international relations. Several of the issues are intertwined, and several arise as a result of the impending retirement of the Space Shuttle.

"The Gap" The Space Station was conceived, designed and built with the Shuttle in mind. Its operational strategy, utilization capacity, and philosophy of maintenance and spares were all developed assuming the availability of the Shuttle.

How will U.S. crew be transported to the ISS after Shuttle retirement? The U.S. will depend on Russian launches until a new U.S. spacecraft and human-capable launch vehicle become operational. For several years the U.S. will pay Russia to transport our astronauts to the ISS. Further, under existing international agreements, the U.S. is responsible for transporting astronauts from Canada, Japan, and the European Space Agency to the ISS, so the U.S. will presumably also be paying Russia for their transport. This period is now expected to extend for seven years.

How will the U.S. transport cargo to and from the ISS? The U.S. plans to stop using Russian Progress vehicles for cargo transport in 2011, although this launch vehicle would continue to fulfill Russian needs. The program of record relies on a combination of international and commercial capabilities currently under development. These include the European ATV and Japanese HTV, each of which has flown to the ISS once, and two new commercial capsules which, along with their rockets, Dragon and Cygnus, are still in development. (See Figure 4.2.2-1.)

The potential issues for the Space Station include: (1) none of these cargo carriers has nearly the cargo capacity of the Shuttle; (2) only the Dragon is planned to have a capability to bring cargo (e.g., experiments, failed parts, etc.) back to Earth; (3) two of these systems have flown once successfully, and the other two are untested. ISS resupply will thus depend on a mix of as-yet relatively less mature or unproven systems after the Shuttle is retired. While the diversity of options gives reason to believe that ISS servicing and resupply can be accomplished, there is little assurance that the new vehicles and capsules will be operational on their planned schedules.

Even today, to supply the ISS with more than the basic essentials for a crew of six using the Shuttle is proving to be a challenge. The Committee notes that while the post- Space Shuttle cargo plan may sustain basic ISS operations, it could put the ISS on a somewhat fragile footing in terms of utilization. There is little surge capacity for unforeseen maintenance or logistics needs, and since utilization has been shown to be the first to suffer when funding pressures rise, the projected capacity may prove insufficient to support meaningful ISS utilization.

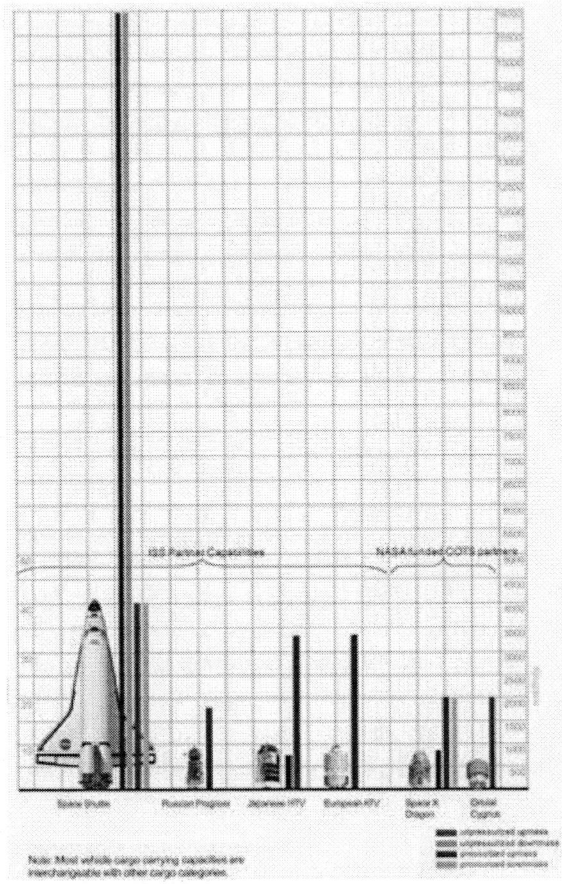

Source: Review of U.S. Human Spaceflight Plans Committee

Figure 4.2.2-1. ISS resupply vehicle payload capacities: Shuttle, Russian Progress, Japanese H-II Transfer Vehicle, European Automated Transfer Vehicle, SpaceX Corp. Dragon and Orbital Sciences Corp. Cygnus. Bars indicate the cargo capacity in kilograms

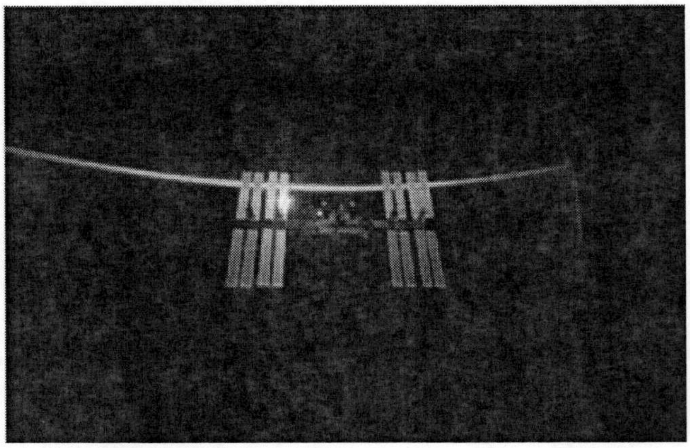

Source: NASA (STS-119 Shuttle mission imagery)

Figure 4.2.2-2. The International Space Station as seen against Earth's horizon

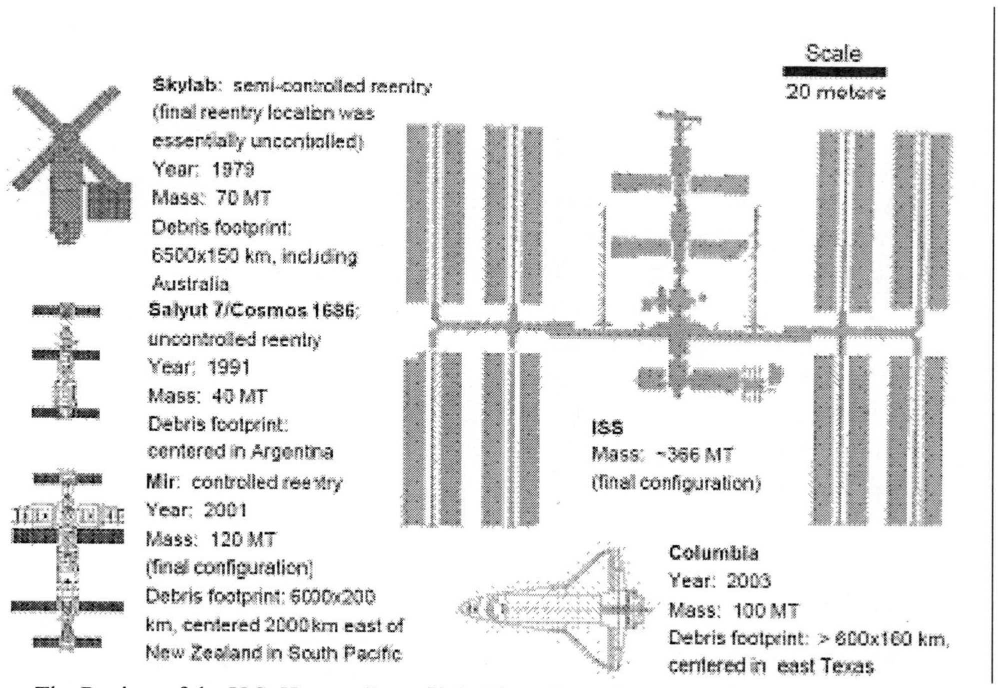

Source: The Review of the U.S. Human Spaceflight Plans Committee

Figure 4.2.2-3. The relative challenge of re-entry of the International Space Station as compared with earlier re-entry/debris events. (Diagrams approximately to scale)

End-of-Life Considerations How and when should the ISS be de-orbited? What should be returned to Earth before ISS de-orbit? Will the "down-mass" capabilities at the time of de-orbit allow significant retrieval of valuable equipment, experiments and facilities? How far in advance of a planned de-orbit should consultations among the international partners take place? These are a few of the issues that must be considered before a de-orbit can be implemented.

Because of its unprecedented size and mass (about 350 mt on orbit), de-orbiting the ISS is not a simple task. (See Figure 4.2.2-3.) There are currently no existing or planned vehicles that could de-orbit the entire ISS in a predictable manner. Thus, either a new de-orbit module would have to be produced and launched to the ISS, or the station would have to be disassembled and the major portions de-orbited individually. The Committee requested an independent assessment of the difficulty of this task, and an estimation of the potential cost. The projected costs are $2 billion or more, depending on the method of de-orbiting required.

The Committee also considered the possibility that the ISS could be operated with minimum U.S. participation, rather than be de-orbited. Preliminary considerations suggest that it would be nearly impossible for the remaining international partners to operate the ISS because of the extreme stress on their smaller budgets, and because U.S. export control requirements would limit the direct support the U.S. could provide to foreign space agencies.

Another alternative would be to "mothball" the ISS in space for later use. In order to assure any future utility, it appears preferable to keep the ISS staffed at a minimum level, similar to that adopted in the early phases of construction. Probabilistic risk assessments find a factor of five increase in probability of loss of the ISS with no crew on board. The need to keep the station occupied would be substantial. There is also a risk that an unoccupied ISS could enter the Earth's atmosphere in an uncontrolled manner, resulting in liability issues and international difficulties for the U.S. In summary, it does not appear that either mothballing the ISS or ending U.S. participation is a viable option, and keeping the Station occupied is very expensive.

The extension of the ISS operations brings its own technical issues. Currently, if a significant part fails on the ISS, this part is returned to Earth and refurbished. Once the Shuttle retires, that will no longer be possible; new parts will have to be procured and lifted to the ISS. To prepare for Shuttle retirement, NASA has begun carrying spare parts up to the ISS—this provisioning is intended to supply the ISS through 2015. If the ISS is extended, additional spares must be procured and the suppliers retained. Further, there are a few parts too large for any of the planned cargo vehicles to lift. In addition, some components of the ISS (e.g., the U.S. laboratory) will reach the end of their certified life in 2015 or shortly thereafter. It is clear to the Committee that if the ISS is to be extended, planning for that should begin immediately.

- 10 International Standard Payload Racks
- Human Research Facility
- Window Observational Research Facility
- Microgravity Science Glovebox

- 8 International Standard Payload Rack
- Biological Experiment Rack
- Fluid Science Experiment Rack
- External Facility supports 10 experiments for space-exposure

- 5 International Standard Payload Racks
- European Drawer Rack
- Fluid Science Laboratory
- Microgravity Science Glovebox
- Biolab
- European Physiology Modules
- European Technology Exposure Facility

- Alpha Magnetic Spectrometer
  - *Anti-He nuclei*
  - *High-energy positrons*
  - *Strangelets*
  - *Cosmic Rays*
  - *Gamma Rays*
  - *Earth-orbiting high-energy particles*
  - *Anti-deuterons*

Source: Review of the U.S. Human Spaceflight Plans Committee

Figure 4.2.2-4. Major research facilities and support capabilities of International Space Station

ISS Utilization and the User Community For the past decade, efforts on the ISS have been directed toward assembly and early operation. Budgetary pressures during construction left little money for utilization. This is still the case. Today, less than 15 percent of NASA's ISS budget is allocated for utilization. As the facility grows, its capacity may not be fully used. (See Figure 4.2.2-4.) Further, the current plan funds ISS utilization at approximately the same level through 2015. At the same time, however, the 2005 NASA Authorization Act designated the U.S. segment of the ISS as a National Laboratory and directed NASA to develop a plan to "increase the utilization of the ISS by other federal entities and the private sector..." It would be difficult if not impossible to realize the potential envisioned in the Congressional language at the current level of utilization.

How well the ISS is exploited depends to a considerable degree on whether its management focuses on utilization. With relatively few U.S. users, it may not have seemed worth restructuring utilization management of a program that was slated for termination after only five years of full operation. In the context of ISS renewal, however, a new management approach that facilitates the use of the ISS by a broad range of scientific, technological, and commercial users is warranted. The Committee believes that an organization is needed to mediate between NASA operations managers and the broad stakeholder community. This could facilitate access to ISS assets by a disparate user community (with widely varying levels of sophistication about spaceflight activities), and could help organize the multiple demands of the users into more unified requirements. Without a mediated dialogue between operations managers and users, it will be difficult to realize operational efficiencies. There are numerous examples of existing organizations that should be examined as possible models.

4.2.3. Scenarios for the future of the ISS

The Committee examined three scenarios for the future of the International Space Station. The first is essentially the program of record; that is, terminate U.S. participation in the ISS at the end of 2015. The second, "steady as you go," renews U.S. participation at the current level to 2020, and assumes that launch vehicle development will proceed at a pace determined by whatever the remaining budget permits. The third enhances U.S. utilization and (possibly) international participation through at least 2020.

Scenario 1: End U.S. participation in the ISS at the end of 2015

The current program of record terminates U.S. participation in the ISS at the end of 2015, and it calls for decommissioning and de-orbiting the ISS by early 2016. This approach is reflected in NASA's current budget projections, though with insufficient funds for de-orbiting. NASA's 2008 Authorization Bill, however, directed the agency to take no steps that preclude extending ISS operations until 2020, and NASA has complied. This scenario constitutes the current program plan. Under this scenario, 15 years of continuous human habitation in space would end in 2015, and be replaced by intermittent sorties, first to low-Earth orbit, and then eventually to the Moon.

While scientific and technological experiments already on the drawing boards may be flown on the ISS in the next five years, it is less likely that new ventures will have enough time to do so. The U. S. starts at a disadvantage in this regard relative to its international partners, since its life science and microgravity science programs are stalled because of budgetary pressures. Congress designated the ISS as a National Laboratory in 2005 to facilitate the development of broad capabilities in science and technology by other government agencies and non-government users, a promising program that is literally just getting off the ground. It is not likely that research will be contemplated or proposed for a facility that may be de-orbited before full value of that research can be realized.

There are also significant international consequences associated with this scenario. By terminating the ISS, the U.S. would voluntarily relinquish its unique area of unchallenged leadership in space. Other nations have been building satellites and launch vehicles and are now constructing human-rated launch vehicles and capsules. But no other nation can match the 20-year U.S. lead in space engineering, construction and operations.

Just as important, by pursuing this option, the U.S. would dismantle a successful multilateral framework for international collaboration—a framework that could be extended in the future for other space projects. By limiting the time that the international partners could realize the return on their investments, the U.S. would be open to the accusation that it is an inconsiderate, if not unreliable, partner. It is unlikely that another international collaboration as broad and deep could be developed soon to replace the current one. New potential partners would be more likely to seek less ambitious bilateral relationships. The Committee's informal consultations with various foreign partner agencies emphasized how important the participation of their astronauts and experiments on the ISS have been to their space activities and to securing public support for their entire space programs. Finally, there is broader domestic and international public opinion that will not unreasonably question whether it is sensible to terminate after five years of full use a project that took 25 years to build.

Our ISS international partners issued a joint statement at a July 2008 Heads of Agency meeting calling for continuation of ISS operations beyond 2015. Russia has declared publicly that it intends to continue operations after 2015, independent of the U.S., if necessary. NASA believes that this is not technically feasible, but the comment is illustrative of the international reaction to the current ISS plan.

The commitment to use commercial vehicles for the ISS resupply is one of the more innovative aspects of the current program. The prospect of an ISS resupply market is already stimulating risk-taking industries to develop new launch vehicles and capsules. However, termination of ISS would abruptly end that market in 2015 after fewer than five years of commercial resupply operations. This may not provide enough opportunity for the new industries to grow to maturity, and in some cases would likely threaten the survival of their efforts in this area.

The Committee estimates that the Ares I vehicle planned to transport humans to low-Earth orbit will not be available until two years after the ISS ceases to operate under the current plan. In this case, there would be several years with no U.S. human spaceflight activity at all. Thus, an achievement gap would exist in addition to the launch capability gap.

Scenario 2: Continue ISS operations at the present level to 2020

Extending ISS operations by five years ameliorates many of the difficulties cited above with the current program (Scenario 1). The U.S. would have a longer time to develop uses of the Station; that is, to rebuild its ISS science program; to develop the ISS National Laboratory; and to provide opportunities to new users. Renewal of the ISS would assure the existing commercial cargo contractors of a more secure market and might also encourage other financial risk-takers to invest. The international partners would have more time to achieve a return on their investments, and the U.S. and its partners would have the opportunity for continuing human activities in space for five more years.

The current level of effort, however, does not allow the ISS to achieve its full potential as a National Laboratory or as a technology testbed. The majority of the funding is devoted to sustaining basic operations and providing transportation (including commercial resupply and crew transport on Soyuz). With utilization only a modest part of the ISS budget, many equipment and experiment racks will remain unfilled, which is the case today.

Extension beyond 2015 does bring new technical issues. These issues include procuring and providing spares and recertifying the components of the ISS. (See Figure 4.2.3-1.) One hidden benefit of work on life extension may be that it provides practical experience with issues that will arise later in missions of exploration beyond low-Earth orbit. As described above, planning for life extension to 2020 would have to begin immediately.

Scenario 3: Enrich the ISS program and extend through 2020

Since ISS utilization accounts for a relatively small portion of the planned budget, a significant enrichment of ISS utilization could be achieved with a relatively modest increase in funding. This is the basis for the Committee's nominal scenario, which is described below. Like Scenario 2, without added funding this scenario would adversely affect the Constellation Program in that it consumes funds that are otherwise planned to be added to that effort.

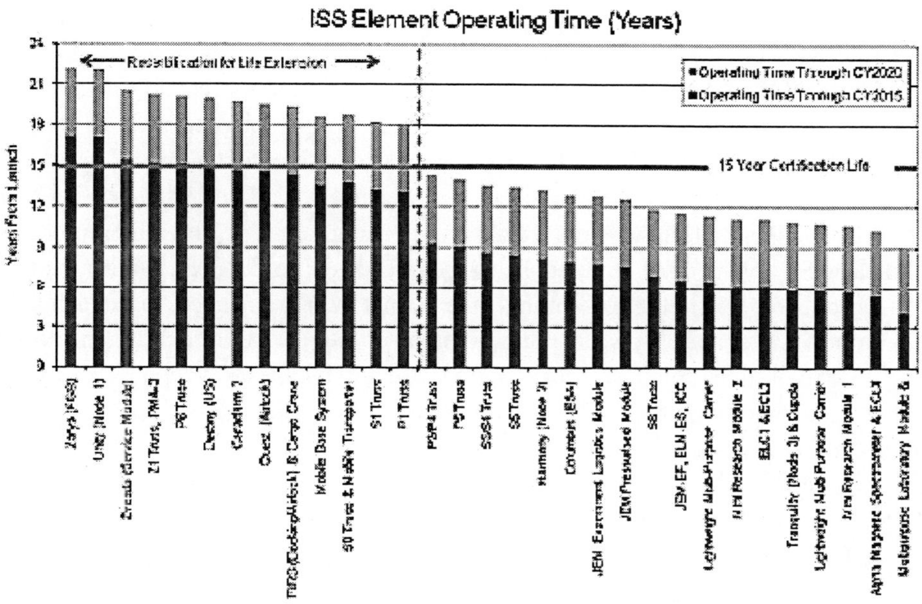

Source: Review of the U.S. Human Spaceflight Plans Committee

Figure 4.2.3-1. Projected lifetime of major International Space Station elements indicating the need for recertifying many elements if the International Space Station is extended to 2020

A much stronger emphasis on utilization will help ameliorate one of the most intractable problems associated with the International Space Station: *Because NASA does not have a compelling vision for how it will use the ISS, many American citizens do not have a clear idea of what it is for.* Further, the absence of funds to support utilization of the Station causes potential users to be skeptical of its overall value. Even if the extension option is adopted, it is not clear whether it will be successful in addressing these concerns. Up to now, the U.S. has focused almost exclusively on building the ISS. Budgetary pressures during construction meant that inadequate attention was paid to how the U.S. would use the facility after it was completed. As one example, the funds originally to be used for research and technology development were reduced. The scientific research community that had hoped to use the ISS has largely been dispersed and will have to be reassembled.

However, there remains the potential to enhance scientific use of the ISS. The National Research Council Space Studies Board has recently initiated a decadal survey of life and microgravity science that will identify key scientific issues and strategies for addressing them. This is the first decadal survey in this area, and it will bring the most modern scientific understanding to bear on what questions may be answered in the decade through 2020. An extended, enriched ISS program will enable more of the scientific opportunities identified by the survey to be captured.

As the nation's newest National Laboratory, the ISS has the potential to further strengthen relationships among NASA, other federal entities and private sector leaders in the pursuit of national priorities for the advancement of science, technology, engineering and mathematics. The ISS National Laboratory should also open new paths for the exploration and economic development of space. The life science research community of the National Institutes of Health and NASA's space station research facilitators recently met for the first time to allow researchers to explore the logistics of flying their experiments on the ISS. Enriching the ISS program would send a strong signal to these potential users.

There is another important use of the ISS that was not considered when the space station was begun in 1984 or redesigned in 1992: to support exploration. The Committee believes that the Space Station can be a valuable testbed for the life support, environmental, and advanced propulsion technologies, among others, that will be needed to send humans on missions farther into space. It also has the potential to help develop operational techniques important to exploration. Such an emphasis has the advantage of keeping the technology development and operational side of NASA involved in ISS utilization.

Among the most compelling considerations supporting this scenario are the opportunities it affords for international partnership. The negotiations to extend the ISS partnership beyond 2016 (which, under the latter two scenarios, should begin soon) offer the U.S. a new opportunity for geopolitical leadership. The ISS partnership can be enriched in a variety of ways: its goals may be enlarged, its membership may be enlarged, or both. By adding aspects of exploration beyond low-Earth orbit to the goals of the ISS partnership, the partners would engage at an early stage with the U.S. in the next grand challenge of space exploration. The ISS agreement itself might serve as the basis for the broader types of agreement that will be appropriate to deep space exploration.

Since the ISS was redesigned in 1992, several nations have developed important new capabilities for robotic and, more recently, human spaceflight. Opening the ISS partnership to new members could engage such emerging space powers with the present international space community, thereby facilitating the exchange of plans, the sharing of financial and intellectual resources, and the same kind of strong working relationships that brought the ISS into being and that sustained it. The Committee's informal consultations with current ISS partner agencies revealed no fundamental reluctance to adding new partners. However, all recommended that the integration of potential new partners proceed after careful discussion and in small steps that could be taken over time.

Findings on the International Space Station (ISS)

Extending the International Space Station

The return on investment to both the United States and our international partners would be significantly enhanced by an extension of ISS life. Not to extend its operation would significantly impair U.S. ability to develop and lead future international spaceflight partnerships.

ISS termination

If the ISS is to be de-orbited in early 2016, negotiations with international partners and operational planning must begin now; additional funds must be added to the budget to accomplish this complex technical task.

ISS utilization

If the life of the ISS is extended, a more robust program of science, human research and technology development would significantly increase the return on investment from the Station and better prepare for human exploration beyond low-Earth orbit. Additional funds would need to be provided for this purpose.

Cargo to and from the ISS

When the Shuttle is retired, the ISS will rely on a mix of commercial and international cargo transports for provisions, resupply, maintenance and utilization. Some of these delivery systems are as yet unproven and could experience delays. While this would not place the ISS itself in jeopardy in the near term, it could put its utilization at risk.

Commercial cargo carriers

NASA's planned transition of much of the ISS cargo resupply to the commercial sector is a positive development. Financial incentives should be added to those suppliers to meet their schedule milestones, as the ISS will be vulnerable until the relevant vehicles have demonstrated their operational capabilities and flight rates.

Management structure for ISS utilization

The benefits of continued operation of the ISS will depend heavily on the extent to which its management focuses on utilization. One possible approach would be to establish an independent organization that mediates between NASA operations managers and the broad stakeholder community of scientific, technological and commercial users.

International partnership in ISS

NASA's international partners value the ISS relationship and U.S. leadership in that relationship. They further view it as a platform for international cooperation in exploration.

4.3. The Constellation Program

In addition to the Shuttle and the ISS, the scope of NASA activities the Committee was directed to examine includes all of the activities within the Exploration Systems Mission Directorate (ESMD). These include the Constellation Program, the name given by NASA to the flight development program for the next phase of human space exploration.

As the Committee assessed the current status and possible future of the Constellation Program, it reviewed the technical, budgetary and schedule challenges that the program faces today. In developing Integrated Options for the nation's human spaceflight program, the Committee established as the baseline what it considers to be an implementable version of the Constellation Program. This baseline case is outlined in more detail in Chapter 6.

The 2004 Vision for Space Exploration established new and ambitious goals for the nation's human spaceflight program. The Constellation Program is NASA's response to that Vision. The Exploration Systems Architecture Study (ESAS) defined the broad architecture for the program in 2005. (See Figure 4.3-1.) The principal program elements include: the Ares I launch vehicle, capable of launching astronauts to low-Earth orbit; the Ares V heavy-lift cargo launch vehicle, to send astronauts and equipment towards the Moon or other destinations beyond low-Earth orbit; the Orion capsule, to carry astronauts to low-Earth orbit and beyond; the Altair lunar lander for descent to the surface of the Moon, and ascent back to lunar orbit for the crew; and surface systems that astronauts will need to explore the lunar surface.

Development of the first two of the elements needed, the Ares I and Orion, is well underway. While development of the Ares V has not been initiated, certain components of the Ares I can be expected to be common with the Ares V. A detailed review by the Committee of the two launch vehicles, the Ares I and Ares V, will be presented in Chapter 5. The Altair lander and the lunar surface systems are still in very early phases of design, and were discussed as part of the Moon First strategy in Chapter 3.

4.3.1. Orion

The remaining principal element, Orion, consists of a spacecraft generally in the shape of the Apollo capsule, a service module and a launch-abort system. Orion is designed to operate in space for up to six months and carry six astronauts, but is currently being configured for ISS support as a four-person vehicle. An upgraded (Block 2) version is an ticipated for travel to the Moon and beyond. Orion performance is constrained by re-entry and landing considerations and is also subject to limitations imposed by Ares I. It has undergone multiple redesign cycles in response to changing requirements.

The Committee examined the design and operations of Orion in some detail. Many concepts are possible for crew-exploration vehicles, and NASA clearly needs a new spacecraft for travel beyond low-Earth orbit. Evidence reviewed by the Committee indicates that the current Orion design will be acceptable for a wide variety of tasks in the human exploration of space. The Committee's greatest concern regarding Orion is its recurring cost. The capsule is five meters in diameter, considerably larger and more massive than previous capsules (e.g., the Apollo capsule), and there is some indication that a smaller and lighter four- person Orion could reduce operations costs. For example, such a configuration might allow landing on land rather than in the ocean, and it might enable simplifications in the (currently large and complex) launch-abort system. In addition, this would also increase launch margin, which could reduce the cost and schedule risk to the Constellation Program. However, a redesign of this magnitude would likely result in well over a year of additional development time and an increase of perhaps a billion dollars in cost. In any case, in order to provide for a sustainable program, every effort should be made to reduce the recurring costs of Orion.

Safety is of course of primary concern in any human-rated system, and Orion, and its companion Ares I launcher, are designed in accordance with NASA's latest human- rating requirements. The design includes an abort capability throughout ascent, as well as requirements to make loss of crew a factor of 10 less likely than at any previous time in human spaceflight. The ability of Ares I to meet these requirements will not be known until it has an established flight record, but it is clearly being designed to a high standard of safety and reliability. A more detailed discussion of human rating is contained in Section 5.3.4.

As part of an independent review, an assessment of the Orion Crew Vehicle "stand alone" development plan was conducted. The assessment focused on critical path elements. It was observed that the Orion development schedule is "back-end loaded," such that designing test articles, conducting tests and producing flight hardware run in parallel, thus creating an extremely high schedule risk. For example, the Program of Record shows only three months between completion of system qualification (June 2014) and the planned Orion 1 launch date (September 2014). A large number of technical risks also add to schedule uncertainty. When compared to historical programs, the most likely delay to the Orion availability approaches 18 months. Additional critical paths exist through ground test and flight test.

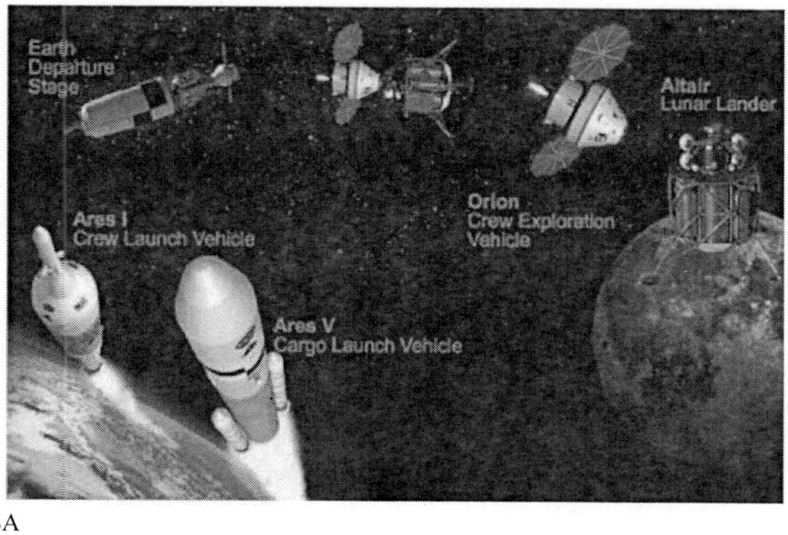

Source: NASA

Figure 4.3-1. The major elements of the Constellation transportation architecture showing sequence of operation

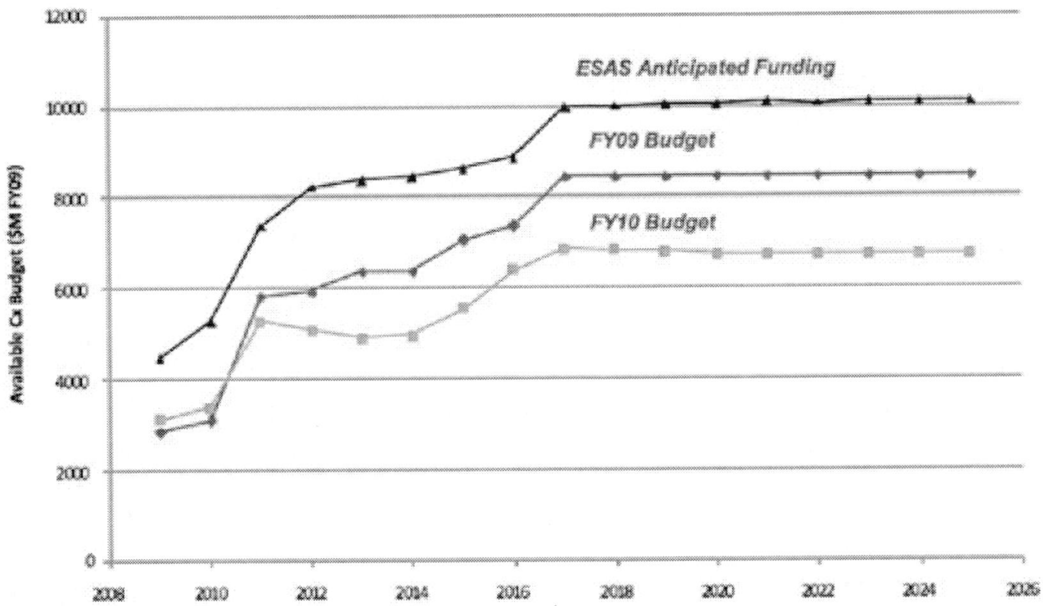

Source: NASA

Figure 4.3.2-1. Constellation Program Funding Profiles

4.3.2. Constellation development

Since Constellation's inception, the program has faced a mismatch between funding and program content. Even when the program was first announced, its timely execution depended on funds becoming available from the retirement of the Space Shuttle (in 2010) and the decommissioning of the ISS (in early 2016). Since those early days, the program's long-term budget outlook has been steadily reduced below the level expected by NASA. As shown in Figure 4.3.2-1, the Exploration Systems Architecture Study of 2005 assumed the availability of a steady-state human spaceflight budget for exploration of about $10 billion per year. In the subsequent FY 2009 and FY 2010 budgets, the long-term projections for funding have decreased. The FY 2010 President's Budget Submittal suggests a steady state funding of about $7 billion per year.

The shorter-term budget situation has had mixed impact on Constellation. The formal first post-ESAS budget was the FY 2007 budget, in which the funds available to Ares I and Orion were significantly lower than those anticipated during the time of ESAS. Subsequently, in the FY 2008 – FY 2010 budgets, the funds anticipated in the out years in the FY 2007 budget were made available to Ares I and Orion. In part this has been achieved by scope changes in other NASA programs.

The budget outlook for the Constellation Program would be even bleaker under some alternate human spaceflight plans. Without additional funding, if the Shuttle manifest extends into 2011 and/or the life of the ISS is extended, there will be even less funding available for Ares I and Orion. Further, as the Shuttle and ISS programs are terminated, a significant percentage of NASA's fixed costs will transition to Constellation. The Committee has found that not all of those costs have been accounted for in the Constellation budget plan. Most major vehicle-development programs face technical challenges as a normal part of the development process, and Constellation is no exception. For example, as the Ares I design has matured, the rocket has grown in weight and various technical issues have emerged. Among these is the high level of vibrations induced by thrust oscillation in the first-stage motor. While significant, these can be considered to be engineering problems, and the Committee expects that they will be solved, just as the developers of Apollo successfully faced challenges such as a capsule fire and an unknown and potentially hazardous landing environment. But finding the solutions to Constellation's technical problems will likely have further impact on the program's cost and schedule.

Differences between the original Constellation program planning budget and the actual implementation budget, coupled with technical problems that have been encountered on the Ares I and Orion programs, have produced the most significant overall impacts to the execution of the Constellation Program. This has resulted, for example, in slipping work on the Ares V and lunar systems well into the future and setting Orion's near-term occupancy at four astronauts.

The original 2005 schedule showed Ares I and Orion available to support the ISS in 2012, only two years after scheduled Shuttle retirement. The current schedule maintained by the Constellation Program now shows that date as 2015, but with a relatively low schedule confidence factor and little schedule slack on the critical path. The Committee commissioned the Aerospace Corporation to perform an independent assessment of the technical, budgetary and schedule risk on the Constellation Program. The study methodology employed for the assessment of cost and schedule is substantially the same as described for the Integrated Options in Figure 6.2.3-1 of Chapter 6. All resulting cost and schedule impacts prepared for the Committee are formulated at the 65-percent confidence level, consistent with the direction NASA has given the Constellation Program.

The results of the analysis in-dicate to the Committee that, under the FY 2010 budget profile, there is likely an additional delay of at least two years, with first launch in 2017, and perhaps as much as four years of delay, with first launch in 2019. This suggests that Ares I and Orion will not reach ISS before the Station's currently planned termination. Assuming a Shuttle retirement sometime in FY 2011, the length of the gap in which the U.S. will have no independent capability to transport astronauts into orbit will be about seven years.

The Constellation Program has identified measures, such as ongoing content reduction, deployment of stimulus funds to address high-risk schedule areas, and program management actions to mitigate major risks, that suggest that the first launch of Ares I and Orion could occur in 2017 if those measures are successful.

The Ares V, still in conceptual design, promises to be an extremely capable rocket—able to lift 160 metric tons of cargo into low-Earth orbit. But its design, too, has experienced growth (and program delays) due to the impact of the development of other elements of Constellation. Under the FY 2010 funding profile, the Committee estimates that Ares V will not be available until the late 2020s. Under the FY 2010 budget, the lunar landing and surface systems will also be delayed by over a decade, indicating that human lunar return could not occur until well into the 2030s.

4.3.3. Importance of technology development

As currently structured, the only broad-based space technology program of NASA is contained within the Exploration Systems Mission Directorate, and is closely tied to the near- and mid-term needs of Constellation. Two recent reports of the National Research Council have examined and made recommendations to NASA on the structure of its future space technology program.

The need for technology development is apparent, and the pursuit of a well-crafted technology program would be very beneficial to the longer-term human spaceflight program. Failure to adequately fund such efforts in the past has reduced the options available today. Further, substantial cost and schedule savings can be achieved by having the needed technologies in place prior to initiating engineering development activities. Almost invariably, to conduct such efforts in parallel is extremely costly.

Based on these considerations, the Committee finds that a robust technology-development program, funded in support of future human spaceflight activities, would not only introduce new opportunities for mission architecture but also enable reduction in the cost of human spaceflight. This will be discussed in Chapter 7.

Findings on Orion and Ares

Orion

The Orion is intended to be a capable crew exploration vehicle, and the current Orion design will be acceptable for a wide variety of tasks in the human exploration of space. The current development is under considerable stress associated with schedule and weight margins. The primary long-term concern of the Committee is the recurring cost of the system.

Ares I

Ares I is intended to be a high reliability launcher. When combined with the Orion and its launch escape system, it is expected to serve as a crew transporter with very high ascent safety. The Ares I is currently dealing with technical problems of a character not remarkable in the design of a complex system – problems that should be resolvable with commensurate cost and schedule impacts. Its ultimate utility is diminished by schedule delays, which cause a mismatch with the programs it is intended to serve.

(Other findings on the Constellation Program launch vehicles are in Chapter 5.)

5. Launch to Low-Earth Orbit and Beyond

Launch to low-Earth orbit is the most energy-intensive and dynamic step in human space exploration. No other single propulsive maneuver, including descent to and ascent from the surfaces of the Moon or Mars, demands higher thrust or more energy or has the high aerodynamic pressure forces than a launch from Earth. Launch is a critical area for spaceflight, and two of the five key questions that guide the future plans for U.S. human spaceflight focus on launch to low-Earth orbit: the delivery of heavy masses to low-Earth orbit and beyond; and the delivery of crew to low-Earth orbit.

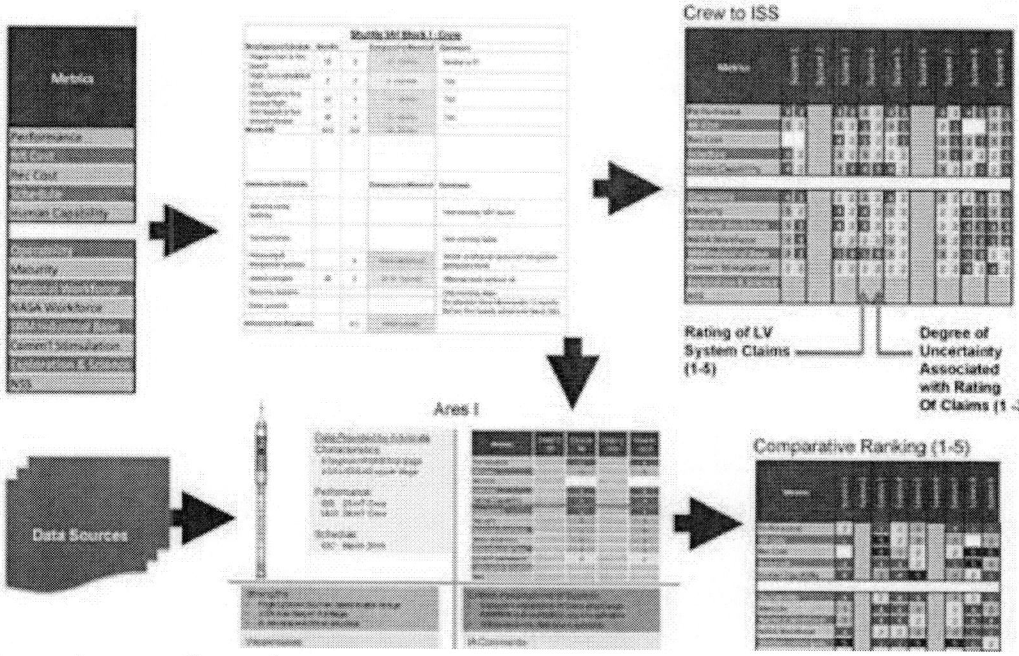

Source: Aerospace Corporation

Figure 5.1-1. Process for assessing metrics for launch vehicle evaluation.

5.1. Evaluation Methodology for Launch Vehicles

Launch vehicles and associated ground infrastructure are key elements of the architectures that support human spaceflight missions. Launch vehicles are generally designed anew or adapted from existing vehicles to support a specified mission or range of missions. The mission definition drives the size, performance, production rates, reliability and safety requirements. This is particularly true for "clean- sheet" (i.e., new) designs. For the adaptation of existing launch vehicles to new missions, greater compromise between the launch vehicle and the mission is often needed in order to execute the adaptation and thus realize the benefits sought. Primary among these benefits is proven safety, cost, reliability and performance.

The Aerospace Corporation performed for the Committee an evaluation of potential launch vehicles. The metrics used in evaluating the various launch vehicle candidates, as shown in the upper left hand corner of Figure 5.1-1, contain the usual cost, performance and schedule parameters, but also include items such as safety, operability, maturity, human rating, workforce implications, impacts on the U.S. industrial base, the development of commercial space, the consequences to national security space, and the impact on exploration and science missions. Some metrics could be evaluated quantitatively, such as cost and schedule, while others required qualitative assessment, such as the impact on as-yet undefined national security space missions.

At a summary level, the assessment process centered around two evaluations. The first was to assess and modify where appropriate the claim for a system as submitted by the provider of that system. The second step was to represent the uncertainty associated with the assessment of each metric for each launch system. This process made it possible to capture cases where claims might be judged to be less than their stated values but with a fair degree of certainty as well as cases where a claim was judged well within historical bounds but significant uncertainty remained about the estimate. It also permitted at least a first-order comparison of existing vehicles with proposed vehicles—including defining the uncertainty in the comparison. A sample summary of these evaluations is shown in the upper right hand corner of Figure 5.1-1.

In analyzing individual launch vehicles the study approach examined approximately 70 lower-level metrics that contributed to the 13 top-level metrics. A summary was created for all relevant launch systems for each mission category as to their ranking relative to the other launch systems capable of supporting that particular mission.

5.2. Heavy Lift to Low-Earth Orbit and beyond

The insertion of heavy payloads from Earth orbit towards their destination is essential for exploration beyond low- Earth orbit, and such systems significantly benefit from heavy lift to low-Earth orbit. The plan of the Constellation Program for the exploration of the Moon envisions launching about 600 metric tons (mt) per year to low-Earth orbit, while exploration along the Flexible Path may require somewhat less launch mass each year. NASA scenarios for the exploration of Mars will have comparable annual requirements. In the three years of lunar surface exploration during Apollo, which had a less capable lunar surface infrastructure than is currently planned, NASA launched over 250 mt per year. As a point of comparison, the ISS, assembled over the last decade by the Shuttle, has a mass of about 350 mt. Thus in the era of exploration beyond low-Earth orbit, we will launch to low-Earth orbit a mass comparable to that of the entire ISS every year.

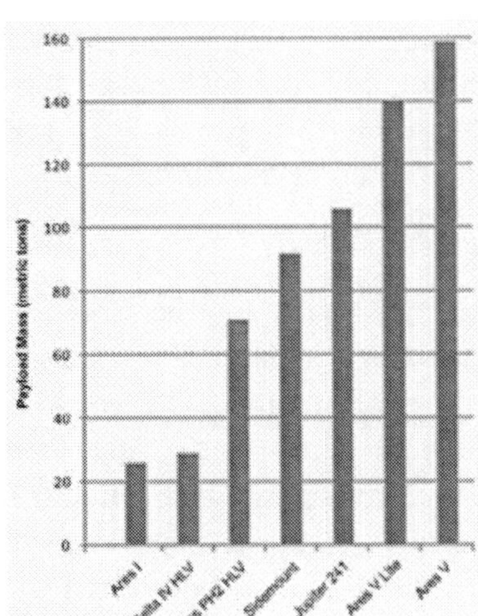

Source: Review of U.S. Human Spaceflight Plans Committee

Figure 5.2-1. Approximate payload mass launched by various launch vehicles to a 28.5 degree inclination low-Earth orbit

Launch Vehicle Family			Launch Mass to LEO	First Stage of Heavy
NASA Heritage	Ares Family	Ares V + Ares I	160 mt + 25 mt	Advanced RS-68 (LOX/LH) + 5.5 segment SRB
		Ares V Lite	140 mt	Advanced RS-68 (LOX/LH) + 5 segment SRB
	Directly Shuttle Derived Launcher		100 -110 mt	SSME (LOX/LH) + 4 segment SRB
EELV Heritage "Super Heavy"			75 mt	RD 180 (LOX/RP-1)

Source: Review of U.S. Human Spaceflight Committee

Figure 5.2-2. Heavy Launch Vehicle Options indicating the approximate launch mass to low-Earth orbit, and the engines of the first stage (LOX is liquid oxygen, LH is liquid hydrogen, RP-1 is a hydrocarbon fuel, SRB is a Solid Rocket Booster derived from the Shuttle system. See text for discussion of engines.)

The key decision for heavy lift to low-Earth orbit is: on what should the next heavy-lift launch vehicle be based? The Committee examined five candidates for heavy lift, with their estimated launch mass to low-Earth orbit shown in the right-most five bars of Figure 5.2-1. In the end, to simplify the considerations, the Committee treated the five launchers in the four classes summarized in Figure 5.2-2: the currently planned Ares I + Ares V architecture; the Ares V Lite (used in a dual mode for lunar missions): a Shuttle-derived vehicle; and a "super-heavy" launcher derived from Evolved Expendable Launch Vehicle (EELV) heritage. The Shuttle- derived class was used to represent both in-line and side- mount vehicles, each of which will be discussed in more detail below.

5.2.1. The need for heavy lift

First, the Committee examined the question: do we need a heavy-lift capability? While it is obvious that the ability to inject massive spacecraft away from low-Earth orbit is vital for exploration, there is some question as to the smallest practical size of the launcher that will be used to carry cargo to low-Earth orbit. The Committee reviewed the issue of whether exploration beyond low-Earth orbit will require a "super heavy-lift" launch vehicle (i.e., larger than the current "heavy" EELVs, whose mass to low-Earth orbit is in the 20-25 mt range), and concluded that it will. However, the rationale for this decision is subtler than usually thought, and hinges on three factors: the size and mass capability of the launcher and of the entire U.S. launch capacity; in-space refueling capability; and the launch reliability expected for a given mission.

No one knows for certain the mass or dimensions of the largest piece of hardware that will be required for future exploration missions. It will likely be larger than 25 metric tons (mt) in mass, and may be larger than the approximately five-meterdiameter fairing of the largest current launchers. The largest single element in the current NASA plans that will be launched to low-Earth orbit is the Earth Departure Stage (EDS). Not counting its approximately 60 mt of payload, the EDS arrives in low-Earth orbit on a standard lunar mission with a mass of about 119 mt, of which about 94 mt is fuel, and only 25 mt is dry mass. In the absence of in-space refueling, the U.S. human spaceflight program will require a heavy-lift launcher of significantly greater than 25 mt capability to launch the EDS and its fuel.

However the picture changes significantly if in-space refueling is used. All of the heavy-lift vehicles listed in Figure 5.2-2 use an EDS to lift the specified payload mass to low-Earth orbit. In the conventional scheme, the EDS burns some of its fuel on the way to orbit, and it arrives in low-Earth orbit partially full. The remainder of the fuel is expended in injecting the payload toward its destination beyond low-Earth orbit. The alternative is to refuel the EDS in low-Earth orbit from either a dedicated tanker or a fuel depot. This allows more mass to be injected from the Earth with a given EDS. Studies commissioned by the Committee found that in-space refueling could increase by at least *two to three times* the injection capability from low-Earth orbit of a launcher system, and in some cases more.

Thus, an in-space refueling capability would make larger super-heavy lift vehicles even more capable, and would enable smaller ones to inject from low-Earth orbit a mass comparable to what larger launchers can do without in- space refueling. (See Figure 5.2.1-1.) In fact, the larger elements launched to low-Earth orbit tend to be propulsion stages, and these are usually about 80 percent fuel by mass. If there were the capability to fuel propulsion stages in space, the single-largest mass launched would be considerably less than in the absence of in-space refueling. The mass that must be launched to low-Earth orbit in the current NASA plan, without its fuel on board, is in the range of 25 to 40 mt, setting a notional lower limit on the size of the super heavy-lift launch vehicle if refueling is available.

As an additional benefit of in-space refueling, the potential government-guaranteed market for fuel in low-Earth orbit would create a stimulus to the commercial launch industry beyond the current ISS commercial cargo-services market.

<table>
<tr><th colspan="3" rowspan="2">Launch Vehicle Family</th><th colspan="2">Launch Mass from LEO toward the Moon (TLI)</th></tr>
<tr><th>No in-space refueling</th><th>With in-space refueling (minimum)</th></tr>
<tr><td rowspan="3">NASA Heritage</td><td rowspan="2">Ares Family</td><td>Ares V + Ares I</td><td>63 mt single launch, 71 mt with Ares I</td><td>~130 mt single launch, ~150 mt with Ares I</td></tr>
<tr><td>Ares V Lite</td><td>55 mt</td><td>~120 mt</td></tr>
<tr><td colspan="2">Directly Shuttle Derived Launcher</td><td>~ 35 mt</td><td>~75 mt</td></tr>
<tr><td colspan="3">EELV Heritage "Super Heavy"</td><td>~ 26 mt</td><td>~55 mt</td></tr>
</table>

Source: Review of U.S. Human Spaceflight Committee

Figure 5.2.1-1. Mass injected away from LEO towards the Moon (on Trans-Lunar Injection) with and without in-space refueling

Source: Boeing

Figure 5.2.1-2. Artist's concept of a fuel depot servicing an Earth Departure Stage.

The Committee examined the current concepts for in-space refueling. There are essentially two. In the simpler one, a single tanker performs a rendezvous and docking with the EDS on orbit, transfers fuel and separates, much like an airborne tanker refuels an aircraft. In a more evolved concept, many tankers rendezvous and transfer fuel to an in-space depot. (See Figure 5.2.1-2.) Then at a later time, the EDS docks with the depot, fuels, and departs Earth orbit. The Committee found both of these concepts feasible with current technology, but in need of significant further engineering development and in-space demonstration before they could be included in a baseline design. This would require engineering effort, and at some development investment, long-term life-cycle savings may be obtained.

The concept of in-space refueling introduces the idea of critical launches and less critical launches in any given mission. Using the lunar mission with crew as a reference, the critical launches would carry the Orion, Altair and EDS to low- Earth orbit. Depending on launch vehicle capacity this could be accomplished on one flight (as it was in Apollo), two, or even three launches. Less critical missions would be the ones that bring fuel to low-Earth orbit. The Committee commissioned a detailed analysis of the reliability of missions that would require multiple launches of critical and less critical payloads. It found that achieving reasonable probability of mission success requires either 90+ days of on-orbit life for the EDS, or a depot, and that at most three critical launches should be employed. Since it is very constraining to balance mission components to always partition equally between launches, this strongly favors a minimum heavy-lift capacity of roughly 50 mt that allows the flexibility to lift two "dry" exploration elements on a single launch.

Another way to view the requirements of heavy lift is to consider the recurring cost to NASA of using a significant fraction of the yearly existing and planned U.S. launch capability after the Shuttle retires. At reasonable production rates of the existing EELV heavy launch vehicles, mid-size EELVs, new commercial vehicles, and the Ares I, much if not all of the excess capability that exists in the U.S. production system would be used launching 400 to 600 mt to low-Earth orbit, and it would be an expensive way to accomplish this.

The Committee finds that exploration would benefit from the development of a heavy-lift capability to enable voyages beyond low-Earth orbit. This might be supplemented by the development of an in-space refueling capability. In-space refueling has great potential benefits, but needs development and demonstration before being incorporated into a baseline design.

Using a launch system with more than three critical launches begins to cause unacceptably low mission launch reliability. Therefore a prudent strategy would be to use launch vehicles that allow the completion of a lunar mission with no more than three launches without refueling. This would imply a launch mass to low-Earth orbit of at least 65 to 70 mt based on current NASA lunar plans. Vehicles in the range up to about 100 mt will require in-space refueling for more demanding missions. Vehicle above this launch capability will be enhanced by in-space refueling, but will not require it. When in-space refueling is developed, any of these launchers will become more capable.

The development of such a heavy-lift vehicle would have other benefits. It would allow large scientific observatories to be launched, potentially enabling them to have optics larger than the current five-meter fairing sizes will allow. More capable deep-space science missions could be mounted, allowing faster or more extensive exploration of the outer solar system. Heavy lift may also provide benefit in national security space applications.

5.2.2. The choices for heavy lift

The Committee examined the four choices for heavy lift outlined in Figure 5.2-1 and Figure 5.2-2, all of which pose different heritage, capability, maturity and organizational ramifications. They will be discussed with reference to the use on a typical lunar mission and Flexible Path mission.

Ares V

The Ares V is used with a human rated Ares I for lunar missions – the so-called 1.5 launch architecture. The Ares I launches the Orion, which docks in low-Earth orbit with the Altair lander and EDS launched on the Ares V. The Orion and Altair departs towards the Moon, propelled by the EDS. (See Figure 3.4.2-1.) The Ares I rocket is currently under development by the Constellation program, and has a great deal of commonality with the Ares V. This version of the Ares V (in contrast with the Ares V Lite discussed below) is the most capable of the launch vehicle alternatives under study, with a payload to low-Earth orbit of about 160 mt. (See Figure 5.2.2-1.) When used in conjunction with the Ares I, the combined payload to low- Earth orbit is about 185 mt. With an appropriately designed lunar lander, one Ares I and one Ares V land about 2 mt of cargo on the lunar surface on a human mission, and about 14 mt reaches the lunar surface on a cargo-only mission with a single Ares V launch.

Both the Ares I and Ares V use Solid Rocket Boosters (SRBs) – a five-segment SRB in the Ares I, and a five-anda-half segment SRB in the Ares V. The Ares V first stage uses six engines from the RS-68 family. The engines are mounted on the bottom of a 10-meter diameter tank. The second or Earth Departure Stage is based on the J2-X engine, which the Ares V shares in common with the Ares I. The advanced RS-68 rocket engine of the core is a modification of the RS-68 engines used on the existing Delta IV launch vehicles. Certain changes to the RS-68 for use in the Ares family are anticipated. These include upgrades to reduce hydrogen flow at startup, and for extended operation in the more aggressive Ares V thermal environment. The J-2X rocket engine is a modification of the J-2 engines used on the Saturn V program. The use of the Ares I as a means of crew transport to low-Earth orbit will be discussed in Section 5.3.

Ares V Lite

The Ares V Lite is a slightly lower performance variant of the Ares V, with a low-Earth orbit payload of about 140 mt, but with the same essential configuration as the Ares V. However, in this option it would be human rated, as it is used for crew launch with the Orion. It uses five-segment SRBs (already under development for the Ares I), and five core engines, a derivative of the RS-68, the engine used on the Delta IV Heavy EELV. It would use a slight variant of the same EDS as the Ares V, with the same J2-X engine. For lunar missions, the Ares V Lite is used in the "dual mode." The Orion and Altair are launched on separate Ares V flights, and they dock either in Earth or Moon orbit, depending on the mission mode eventually chosen.

With a combined payload to low-Earth orbit for the dual launches of about 280 mt, this architecture would enjoy considerable payload margin that could provide significant enhancement in mission robustness for lunar missions. With an appropriately designed lunar lander, this system lands about 7 mt of cargo on the lunar surface on a crewed mission. With a single Ares V Lite launch, the same lander can deliver about 14 mt on a cargo-only mission.

When used on Flexible Path missions, a single Ares V Lite and EDS, in combination with the Orion and an in-space habitat, are able to support a visit to the Lagrange points without refueling. In order to reach near-Earth objects and beyond, in-space refueling (or alternatively multiple Ares V Lite launches) is necessary.

Shuttle-derived launchers

The Committee examined the Shuttle-derived family, consisting of in-line and side-mount vehicles substantially derived from the Shuttle. These are all characterized by four-segment solid rocket boosters, Space Shuttle Main Engines (or their RS-25E expendable derivatives), and 8.3-meter-diameter external tanks, as used on the Space Shuttle. This class actually comprises a family of possible vehicles.

Source: NASA

Figure 5.2.2-1. Ares V.

Shuttle derived side-mount launcher

On one end of the spectrum is the side-mount launcher that is most directly derived from the Shuttle. (See Figure 5.2.2-2.) It replaces the Shuttle orbiter with an expendable side-mounted payload carrier that contains both a clamshell-type payload shroud and a propulsion system consisting of three Space Shuttle Main Engines (SSME). It uses the exact same tank and SRB configuration as the Shuttle. It would likely carry an EDS with a J2-X engine internal to the payload bay. This would be the smallest development effort possible within the family. Depending on details of design, a side-mount launcher with an EDS can lift 90 to slightly more than 100 mt to low- Earth orbit.

Shuttle-derived inline launcher

At the other end of the spectrum is an inline vehicle, such as the Jupiter 241, with four SSMEs mounted at the bottom of a redesigned tank/ thrust structure, and with an EDS atop the tank. (See Figure 5.2.2-3.) Once the existing stock of SSME engines is depleted, this configuration will use an expendable version of the SSME, the RS-25E. The upper stage uses liquid oxygen and liquid hydrogen as propellants, with a single J-2X rocket engine. The Committee evaluated the smaller Jupiter 130 that does not have an EDS, but used the more capable Jupiter 241 as the basis of comparison, which has a launch capacity in the range of 100 to 110 mt to low-Earth orbit.

In a lunar exploration scenario, it is assumed that three Shuttle-derived launchers of a nominal 110 mt capability would be used for a crewed mission. A single launcher would be used for lunar cargo missions. When used in conjunction with a lunar lander designed for this size vehicle, this system lands about 5 mt of cargo on a crew mission. A single-launch cargo mission lands less than 5 mt of cargo on the lunar surface. However, a single launch cargo mission, enhanced by the use of in-space refueling, increases the cargo mass landed on the lunar surface to more than 20 mt. For Flexible Path missions, two Shuttle-derived launchers, combined with in-space refueling of an EDS, are required to propel the Orion and in-space habitat to the Lagrange points, near-Earth objects and beyond.

While the Committee did not examine the technical trade between the side-mount and inline variants in detail, it observes that the side-mount variant is considered an inherently less safe arrangement if crew are to be carried, and is more limited in its growth potential.

Source: NASA

Figure 5.2.2-2. Shuttle-Derived Sidemount Launcher

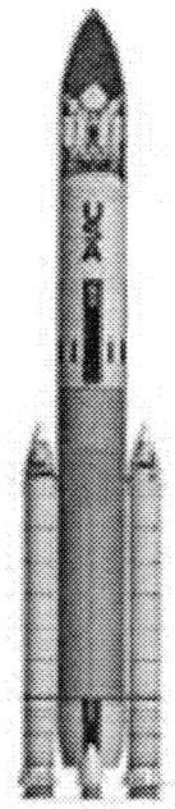

Source: NASA

Figure 5.2.2-3. Jupiter 241

EELV heritage super-heavy

The EELV heritage super-heavy launchers represent a potential family of vehicles derived from the current Evolved Expendable Launch Vehicles heritage. They are distinguished technically from the NASA heritage vehicle by their use of liquid booster (rather than the solid rocket boosters) and secondarily by a hydrocarbon (RP-1) fueled rocket in the first stage core. (See Figure 5.2.2-4.) The upgraded EELV systems would have a core vehicle that would, by itself, have a launch capability to low-Earth orbit in the range of 30 to 35 mt. Using a "super-heavy" variant that would have a core and two boosters of the same basic design, and when used in conjunction with an upgraded common hydrogen/oxygen upper stage, it is likely to have a maximum payload to low-Earth orbit in the range of 75 mt. This exceeds the nominal minimum for a heavy lifter useful for exploration as defined above. A representative of this category of launchers is the Atlas 5 Phase 2 Heavy.

Source: NASA

Figure 5.2.2-4. EELV Heritage Super Heavy Launcher

The EELV super-heavy uses two RD- 180 rocket engines on each of the core and two boosters. The RD- 180 engine has a long history of successful launches in Russia and in the U.S. on the Atlas V family of launch vehicles. The upper stage uses liquid oxygen and liquid hydrogen as propellant, with four RL10 rocket engines. The RL10 family of engines has a long history of successful launches on programs including the Titan, Atlas and Delta families of vehicles. The Atlas Phase 2 is a proposed follow-on of the EELV program, with larger, five-meter diameter, stages, manufactured using the existing five-meter production facilities that currently produce the Delta IV core stages. When used in conjunction with Flexible Path missions, the EELV-heritage launcher and EDS send the Orion and in-space habitat towards the Lagrange points, near-Earth objects and beyond, with two launches plus in-space refueling.

Developmental considerations

Any of these launcher options will entail a substantial development project. The Committee observes that throughout the history of launcher development, and particularly for the Shuttle, the aim has frequently been to design for ultimate performance, often at the cost of reliability and operational efficiency. In particular, NASA's design culture has repeatedly focused on maximizing performance at minimum development cost, generally resulting in high operational and lifecycle costs. While performance is important in launchers, good performance margins and associated robustness are also desirable, and can lead to lower life-cycle costs. A shift in NASA design culture toward design for minimum life-cycle cost, accompanied by robustness and adequate margins, will allow NASA programs to be more sustainable.

There is one additional consideration regarding heavy lift capability. In all missions beyond low-Earth orbit, there will be a need for one or two additional propulsive maneuvers far from Earth. For example, in visiting the Moon, a burn is necessary to enter lunar orbit, and another to leave. When visiting a near-Earth object, a burn is necessary to decelerate to rendezvous with the object, and then a second to return to Earth. Exploration will require a long-duration in-space restartable stage. This would become a building block of exploration propulsion systems, potentially including the lunar descent stage.

Findings on Launch to Low-Earth Orbit and beyond

The need for heavy-lift

A heavy-lift launch capability to low-Earth orbit, combined with the ability to inject heavy payloads away from the Earth, is beneficial to exploration, and will also be useful to the national security space and scientific communities. The Committee reviewed the Ares family of launchers, Shuttle-derived vehicles, and launchers derived from the EELV family. Each approach has advantages and disadvantages, trading capability, lifecycle costs, operational complexity and the "way of doing business" within the program and NASA.

In-Space Refueling

The ability to add fuel to an Earth- departure stage, either from in-space docking with a tanker or from a depot, is of significant potential benefit to the in- space transportation system beyond low-Earth orbit. The technology for in-space refueling is available, but a further development and demonstration program is required. Therefore a prudent approach is to develop a heavy-lift launch system with sufficient capabilities for early missions, which would later be enhanced by in-space refueling when it becomes available.

Sustainability of operations of U.S. launch systems

NASA's design culture emphasizes maximizing performance at minimum development cost, repeatedly resulting in high operational and lifecycle costs. A shift in NASA design culture toward design for minimum discounted life-cycle cost, accompanied by robustness and adequate margins, will allow NASA programs to be more sustainable.

In-space propulsion

For almost all foreseeable missions beyond low-Earth orbit, there is a need for one or two propulsive maneuvers, often after weeks or months in space. Efficient engines and stages with high-reliability restart capability will need to be developed.

5.3. Crew Launch to Low-Earth Orbit

Among the most safety critical aspects of human spaceflight is the delivery and return of crew to and from low-Earth orbit. The fourth key question examined by the Committee is: how should U.S. crew be transported to low-Earth orbit? There are two choices for transporting U.S. crews to and from low-Earth orbit that emerged from the work of the Committee: a government-provided and operated system, and a commercially provided crew-delivery service. This discussion assumes that the Orion vehicle will be the primary U.S. capsule for crew transportation beyond low-Earth orbit and re-entry into the Earth's atmosphere upon return from those voyages.

5.3.1. Ares I plus Orion: Government-provided crew to low-Earth orbit

The current NASA plan for crew transport to low-Earth orbit—comprising the Ares I launch vehicle and the Orion crew capsule—was selected in 2005 as part of the ESAS study based in part on the anticipated availability and projected crew safety considerations of the Ares I and Orion. At the time of ESAS, estimates showed that Ares I and Orion would be available for crew transport service to the ISS by 2012. The date projected by the Constellation program is now 2015. As the plan evolved after 2005, the Ares I developed increasing commonality with the Ares V, providing architectural synergy and reducing development costs of the family.

Ares I

The Ares I launch vehicle currently consists of a single five-segment solid rocket booster as the first stage and a liquid-fueled upper stage. (See Figure 5.3.1-1.) The five- segment SRB is a modification to the existing Space Shuttle SRBs. The upper stage uses liquid oxygen and liquid hydrogen as propellants, with a single J-2X rocket engine. The J-2X rocket engine is a modification of the J-2 engines used on the Saturn V program.

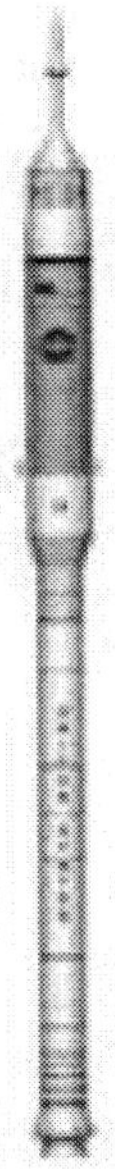

Figure 5.3.1-1. Ares I

In its selection of a crew launch system, ESAS correctly placed a very high premium on crew safety, and the Ares I was selected because of its potential delivering at least ten times the level of crew safety as the current Shuttle. The launch vehicle configuration is one that best allows for crew escape in the event of a launch failure. The capsule is mounted at the top of the stack, and has an independent launch escape sys-ascent.

Under the budget profile NASA leadership anticipated in 2005, estimates showed that the Ares I could be developed by the early to mid part of the decade, and the Ares V could be developed by the late 2010s. It was thought that the Ares I would have lower operating cost when visiting the ISS than other alternatives, and would produce a lower operating cost of the entire system when joint operations of the Ares I and Ares V were begun.

Additionally the development approach of engaging many NASA employees in the design and testing of the Ares I would allow the NASA workforce, which has not developed a new rocket for over 20 years, to gain experience on the relatively simpler Ares I rocket before beginning the development of the more complex Ares V.

5.3.2. Alternatives to government-provided crew access to low-Earth orbit

The Committee considered several alternatives to Ares I and Orion, including:

- A longer-term reliance on international (currently Russian) crew transport services
- The human rating of an existing EELV for launching the Orion
- The development of commercial crew transport services
- The use of a heavy-lift vehicle to launch the Orion

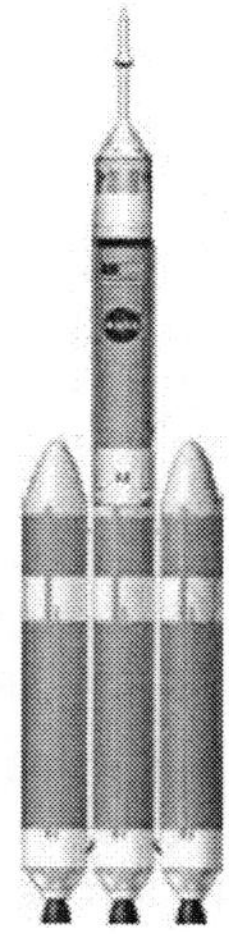

Source: NASA

Figure 5.3.2-1. Delta IV – HLV

While the Committee found interim reliance on international crew transport services acceptable, it also found that an important part of sustained U.S. leadership in space is the operation of our own domestic crew launch capability. This closed out the first alternative. The Committee next examined a NASA-commissioned study by the Aerospace Corporation on the feasibility and cost of human-rating an EELV, the Delta IV Heavy, for use as the launcher for Orion.

Delta IV HLV

The Delta IV Heavy Launch Vehicle consists of two liquid-fueled strap-on boosters, a liquid-fueled first stage, and a liquid-fueled upper stage. The two strap-ons and the core stage are very similar and use liquid oxygen and liquid hydrogen as propellants and a single RS-68-family rocket engine on each of the three stages. The independent study found that launch of Orion to low-Earth orbit did not IV HLV is a variant of one of the EELVs that has launched successfully many times.

While launch of the Orion on the Delta IV HLV was found to be technically feasible, it requires some modification of the current launcher, and was comparable in cost and schedule to simply continuing with the development of the Ares I. When the Committee factored in the carrying cost of the NASA infrastructure that would be maintained if any NASA-heritage heavy launcher would eventually be developed (Ares V in any variant or a more directly Shuttle-derived heavy launcher), any cost savings that might have occurred due to using an EELV to launch the Orion were lost. Using the EELV for launch of Orion would only make sense if it were coupled with the development of an EELV-heritage super-heavy vehicle for cargo launch. Except in this case, this analysis closed out the second option.

5.3.3. Commercial services to transport crew to low-Earth orbit

Having eliminated the long-term international supply option and the EELV option for all but the EELV-heritage super-heavy choice for heavy lift, the remaining possible choices, besides Ares I, were to utilize commercial crew services or use the heavy-lift vehicle as a crew launcher. As the nation moves from the complex, reusable Shuttle back to a simpler, smaller capsule, it is an appropriate time to consider turning this transport function over to the commercial sector. There is broad policy support for this approach, from both Congressional legislation and Presidential policy (See Figure iii.), and one of the four main charges given to the Committee by the Office of Science & Technology Policy in its Statement of Task was to "Stimulate commercial spaceflight capability." This section considers the technical feasibility of a commercial service, safety issues, financial implications, programmatic risks, and acquisition strategy.

Technical feasibility of commercial transport services for crew

The Committee examined the technical feasibility of utilizing a commercial service to transport crew to low-Earth orbit. First, it is a statement of fact that all of the U.S. crew launch systems built to date have been built *by* industry for NASA. The system under contemplation is not much more complex than a modern Gemini, which was built by U.S. industry over 40 years ago. It would consist of a three- or four-person crew taxi, launched on a rocket with a launch escape system. It would have an on-orbit life independent of the ISS of only weeks, but potentially be storable at the ISS for months. Such a vehicle would re-enter the Earth's atmosphere from the speed of orbital flight, rather than the significantly higher speed for which Orion is designed. Its smaller size makes possible the option of landing on land, potentially reducing operations cost when compared to a sea landing.

Recently, several aerospace companies began developing new rockets and on-orbit vehicles as part of the commercial cargo delivery program. Several other U.S. companies are contemplating orbital passenger flight. There is little doubt that the U.S. aerospace industry, from historical builders of human spacecraft to the new entrants, has the technical capability to build and operate a crew taxi to low-Earth orbit.

NASA's role in safety and mission assurance

The Committee treated the safety of crew vehicles as the *sine qua non* of the human spaceflight program, and would not suggest that a commercial service be provided for transportation of NASA crew if NASA could not be convinced that it was substantially safe. The critical question is: can a simple capsule with a launch escape system, operating on a high-reliability liquid booster, be made safer than the Shuttle, and comparably as safe as Ares I plus Orion? An important part of this analysis rests on the reliability of the launcher. Thus, commercial crew launchers based on high-reliability vehicles that already have significant flight heritage, or will develop flight experience soon, would be more obvious candidates as a crew launcher. Ares I has a heritage that traces to the use of the SRBs on the Shuttle, but other potential crew launchers can also trace their lineage to significant flight heritage.

Given the history of human spaceflight, putting commercial crew transport to space in the critical path of any scenario represents a major shift in policy. As will be discussed in Section 5.4.2., the Committee reviewed convincing evidence of the value of independent oversight in the mission assurance of launchers, and would envision a strong NASA oversight role in assuring commercial vehicle safety. The challenge of developing a safe and reliable commercial capability for crew transport will require devoting government funds to "buy down" a significant amount of the existing uncertainty. Whatever the particulars of this risk removal process, it will take an appreciable period of time and require the application of thorough, independent mission-assurance practices. A critical aspect of this exercise will be confirming the root cause and adequacy of correction of any failures or anomalies encountered in the development test program. Thus, the Committee views any commercial program of crew transport to ISS as involving a strong, independent mission assurance role for NASA.

The Committee identified elements of a plan that would lead to the creation of a commercial service for crew transport, building on NASA incentives and guarantees. This included an assessment of the financial aspects and benefits of commercial crew services, the programmatic risks of relying on commercial crew services and potential mitigation strategies, and an approach to engaging the commercial community in this program.

Financial aspects of the commercial crew services

The Committee engaged in a two-step process for assessing the potential financial benefit of commercial services for crew transport. This involved both estimating the cost to develop and operate the system, and then determining what fraction of this cost NASA would likely have to provide as an incentive to industry to enter into this venture.

During its fact-finding process, the Committee received proprietary information from five different companies interested in the provision of commercial crew transportation services to low-Earth orbit. These included large and small companies, some of which have previously developed crew systems for NASA. The Committee also received input from prospective customers stating that there is a market for commercial crew transportation to low-Earth orbit for non-NASA purposes if the price is low enough and safety robust enough, and from prospective providers stating that it is technically possible to provide a commercially viable price on a marginal cost basis, given a developed system. None of the input suggested that at the price obtainable for a capsule-plus-expendable-launchvehicle system, the market was sufficient to provide a return on the investment of the initial capsule development. In other words, if a capsule is developed that meets commercial needs, there will be customers to share operating costs with NASA, but unless NASA creates significant incentives for the development of the capsule, the service is unlikely to be developed on a purely commercial basis.

The Committee then estimated the cost to NASA of creating an incentive for industry to develop the commercial transport capability for crew. This would probably be a significant fraction, but not the entirety of the cost of such a development. Given a properly structured procurement, estimates the Committee received from potential providers for the price of reaching initial demonstration flight of a crew-taxi capsule ranged from $300 million to $1.5 billion. For estimating purposes, the Committee assumed that three contracts were initiated, and one competitor subsequently dropped out, suggesting an expected cost to NASA of between $2 billion and $2.5 billion. In addition, the Committee believes that if a commercial crew program is pursued, NASA should make available to bidders a suitable version of an existing booster with a demonstrated track record of successful flight, adding to the program cost. The best preliminary estimate of the Committee was about a $3 billion program for the fraction of the design, development, test, and evaluation (DDT&E) effort that would be borne by NASA. After multiplying by the historical growth factors and other multipliers associated with 65 percent confidence estimating (as will be discussed in Section 6.3), the cost carried in the Committee's final estimate of the cost of the program to NASA is about $5 billion.

Comparing the scope of providing a commercial crew capability to the cost of historical programs offers a sanity check. In the existing COTS A-C contracts, two commercial suppliers have received or invested about $400-$500 million for the development of a new launch vehicle and unmanned spacecraft. Gemini is the closest historical program in scope to the envisioned commercial crew taxi. In about four years in the early- to mid-1960s, NASA and industry human-rated the Titan II (which required 39 months), and designed and tested a capsule. In GDP-inflator-corrected FY 2009 dollars, the DDT&E cost of this program was about $2 .5-3 billion, depending on the accounting for test flights. These two comparatives tend to support the estimate that the program can be viable with a $5 billion stimulus from NASA.

The Committee considered several other factors that would support this estimate of the incentive cost to NASA. If this is to be a commercial venture, at least some commercial capital must be at risk. Alternate sources of capital, including private and corporate investment, would be expected. Next, the Committee considered the cost associated with the development of a relatively simpler launcher and capsule designed only as a low-Earth orbit crew taxi in comparison with those associated with the far more capable Ares I and Orion. Additionally, the Committee heard many argue that economic efficiencies could be found by striking a better balance between the legitimate need for a NASA quality assurance and safety process on one hand, and allowing industry to execute design and development efficiently on the other.

Significantly, the Committee considered the fact that some development costs, and a larger fraction of operating costs of a commercial crew service to low-Earth orbit could be amortized over other markets and customers. This is more obvious for the launcher, which potentially could also be used for the existing markets of ISS cargo to low-Earth orbit, science and national security space satellite missions, and commercial satellite launches. In the future the commercial booster used for crew might also launch fuel to low-Earth orbit for in-space refueling, and it might carry additional non-NASA crew flights. The non-NASA markets and customers for the capsule are less easily quantified. It is possible that other governments would procure crew launch to ISS from a U.S. commercial provider, and that private travel to low-Earth orbit could be more common by the latter part of the decade. Note that if there were only one non-NASA flight of this system per year, it would reduce the NASA share of the fixed recurring cost by 33 percent.

It was estimated by the Committee that under the "less- constrained budget" to be discussed in Chapter 6, the commercial crew launch service could be in place by 2016. Estimates from providers ranged from three years to five years from the present. Assuming a year for program re-alignment, this would produce a start in early FY 2011. Using the upper end of the estimated range, a capability in 2016 could be estimated with reasonable confidence.

Programmatic risks of commercial crew to low-Earth orbit and potential mitigations

The Committee recognizes that the development of commercial services to transport crew come with significant programmatic risks.

Among these are that the development of this capability will distract current potential providers from the near-term goal of successfully developing commercial cargo capability. Second, the commercial community may fail to deliver a crew capability in mid-program, and the task would revert to NASA. This could be caused by either a technical failure or a business failure—a failure to obtain financing, changes in markets or key suppliers, re-alignment of business priorities, or another non-technical reason. Either type of failure would require NASA intervention, and the possibility that NASA would either have to operate the system, or fall back to an alternative.

While there are many potential benefits of commercial services that transport crew to low-Earth orbit, there are simply too many risks at the present time not to have a viable fallback option for risk mitigation. The Committee contemplated several alternatives, including continuing to rely on international providers (likely available, but not consistent with the long-term need for U.S. access as part of its leadership in space), and continuing the Ares I program in parallel (prohibitive in terms of cost). The Committee also considered the possibility of putting the Ares I program on "warm hold," ready for a possible restart, while continuing the development of the five-segment SRB and the J2-X, which are common to the Ares V. In the end, the Committee thought that the most cost-effective fallback option that would move NASA most rapidly toward exploration is to continue to develop the Orion, and move as quickly as possible to the development of a human-ratable heavy lift vehicle. (See the discussion in Section 5.3.4 on human rating.) The first stage of any of the heavy-lift launchers under consideration would be more than capable of launching an Orion to low-Earth orbit. In the best case, the heavy-lift vehicles themselves would not be available until the beginning of the 2020s, but the first stage or core could be accelerated, perhaps by a year or two. In this case, the core heavy human-rated launcher would arrive only a few years later than an Ares I under the less-constrained budget scenarios.

The details of the preferred fallback option are best left to NASA, but the question is clear. Assume that emphasis will be placed on building the heavy lifter as quickly as possible, and assume that commercial services for crew transportation to low-Earth orbit will be started in development, and may fail to materialize. What variant of the heavy launcher can be identified that could be developed quickly and at small marginal cost in the future if needed to transport Orion and crew to low-Earth orbit? A desirable feature is that the preparation for the development of this variant would have the minimum impact on the construction of the heavy-lift vehicle itself.

Engaging the commercial community in crew transport services

The potential providers of capsules emphasized that the nature of the acquisition of these services is critical; to be commercially viable, low operating cost is essential, and to obtain that cost, the requirements for the capsule need to be as few as is essential and stable. Several providers gave anecdotal examples where NASA programs suffered from significant "requirements creep," and emphasized the need for a more commercial-type procurement, where any changes to work scope would be matters for mutual negotiation, rather than one-sided impositions by NASA.

The Committee envisions a new competition for this service, in which both large and small companies are invited to participate. Several potential providers should be funded through some initial development milestones measuring tangible progress, and incrementally incentivized. It is crucial to the success of the program that multiple providers be carried through to operational service. It is the pressure of competition that provides the drive for low operating cost. Assurances of a market would need to be offered by the government. By creating a third market for commercial launch services (cargo to ISS, fuel to low-Earth orbit, and crew to low-Earth orbit), it is possible that the efficiencies associated with increased production runs and more frequent operations will appear.

5.3.4. Human rating of launchers

The history of human rating U.S. launch vehicles can be traced to the Atlas and ICBM Titan usage in the Mercury and Gemini programs. The purpose of human rating was, and is, to assure that safety levels are appropriate for human flight. Crew safety was addressed in these earlier programs primarily by adding a crew escape system. The reliability of these launch vehicles was addressed by eliminating known design weaknesses, adding redundancies, providing fault detection systems (to initiate crew escape), and tightening requirements for manufacturing, assembly, systems test and checkout at the launch sites.

The process of human rating launch vehicles is central to the viability of commercial service for crew transport, as well as the option of using the heavy launcher as a backup. The Committee found a progressive new approach to human rating at NASA, reflected in the current human-rating guidelines (NASA Procedural Requirement 8705.2B). These guidelines, applicable to newly developed NASA vehicles, provide for intelligent application of similar and dissimilar redundancy when called for, and appropriate approaches to single-string design when unavoidable. In addition, there is a general set of guidelines (NPR 8715.3) that currently would be applied to NASA personnel operating in non- NASA vehicles.

In view of the complexity and cost of retroactively human rating a vehicle (comparable to a significant fraction of the original cost to develop the vehicle), the Committee suggests that all new NASA-developed vehicles, including heavy-lift launchers, be designed so that they are human-ratable, i.e., they could be reasonably human rated at some point in the future. This is a compromise between human rating them at inception and not human-rating them at all. It preserves the option to human rate in the future at lower cost. NASA would benefit from this approach so that it could use its heavy-lift launcher as a backup crew vehicle with Orion, should the commercial providers fail to deliver for any combination of business and/or technical reasons. Additionally, the criticality of cargo launched on the heavy-lift vehicle would suggest that NASA institute quality control and requirements comparable to human-rating guidelines in any event.

Findings on Crew Launch to Low- Earth Orbit

(Note: a finding on the Ares I is presented in Chapter 4.)

The need for independent U.S. human access to Space:

In the long run, it is important for the U.S. to maintain independent access to low-Earth orbit for its crews. In the future, this might be provided by government, commercial providers, or a combination of the two.

Commercial launch of crew to low-Earth orbit

Commercial services to deliver crew to low-Earth orbit are within reach. While this presents some risk, it could provide an earlier capability at lower initial and life-cycle costs than government could achieve. A new competition with adequate incentives should be open to all U.S. aerospace companies. This would allow NASA to focus on more challenging roles, including human exploration beyond low-Earth orbit based on the continued development of the current or modified Orion spacecraft.

Human rating of launch vehicles

NASA has recently adopted a new, more outcome-based standard for human rating space systems, and has in its policy a more flexible approach for human rating existing or new third-party systems. NASA would be well served by applying these policies to field a set of safe yet efficient capabilities, including the provision that all newly developed, government–funded launch vehicles be readily human-ratable. In this way, if plans change in the future—for example, if the commercial capability of crew transport to low-Earth orbit fails to materialize—NASA would have a backup means of launching crew on heavy lift vehicles.

Safety

Human space exploration is an inherently risky endeavor. NASA should continue to make every reasonable effort to reduce the risks in spaceflight. Design for safety should be the prime but not only criterion in the development of systems and operations.

5.4. Additional Issues in Launcher Selection

5.4.1. Launch vehicle performance and costing

In evaluating the systems described in this chapter, the Committee noted that each has avid proponents, and as such the claimed cost, schedule and performance parameters include varying degrees of aggressiveness. Some of these estimates are close to or within the spread of historical programs, while others are well outside historical bounds. The latter could of course be attributed to fresh new approaches that make the historical databases inapplicable—or they could be attributable to unwarranted optimism. The analysis techniques employed in the assessment sought to differentiate between the two.

The only large potential decrease to the cost of space transportation, absent greatly increased traffic, resides in the adoption of a new paradigm for commercially purchasing highly reliable space transportation services. This approach benefits from commercial best practices embraced by experienced providers of launch systems and new systems and processes offered by young firms. The price of shifting to commercial practices is to decrease the NASA workforce and sacrifice the expertise that has been built up at NASA over the years as the agency has directed and overseen the development of launch systems. This new strategy may eventually restore the total national space launch workforce in terms of expertise and number of workers, but near-term reductions would be expected.

The health and viability of the large solid-rocket motor industrial base rests in part on the choice of future crew transport and heavy-lift cargo launch system designs. Those that are all-liquid obviously present the most negative impact on the SRM industrial base and those that either launch crew with an SRM or support heavy-lift with SRMs provide the most benefit to the SRM industrial base. If the choice is to pursue all-liquid launch systems for both crew and cargo, there is no perceived future need for large segmented SRMs in support of civil space activities.

5.4.2. Reliability

Reliability of launchers is important to the safety of crew, and the success of missions of exploration. The U.S. history of heavy-lift launcher reliability is shown in Figure 5.4.2-1. The historical record can be separated into two classes: those intended for human spaceflight, and those intended for cargo use. Saturn and the Shuttle are in the former class, while the other existing vehicles considered here are in the cargo class.

The most reliable U.S. heavy-lift launch vehicle built to date is the Shuttle, whose launch reliability, demonstrated by flight history, and computed using the Bayesian estimation process (based on a 50-percent confidence level, common for this type of analysis) is 98.7 percent. The Saturn vehicles also exhibited high reliability over their limited flight history. The less reliable Titan HLV only launched cargo, and the Delta IV H is limited to only three flights to date.

During this examination of future space exploration launcsystems, the vehicles considered included those derived from Shuttle, Saturn and Delta IV–Heavy-heritage engines and motors (including Ares I and V and Shuttle derivatives), those derived from the EELV program families of launchers (Atlas V and Delta IV), and new vehicles with limited heritage (Falcon and Taurus II). Historically, vehicles with heritage derived from prior demonstrated systems have shown greater reliability in early usage than newly developed systems. The process of converting an established cargo launcher into a human-rated launcher results in improved reliability, as was demonstrated in the early U.S. human spaceflight programs where modified ICBMs were employed as launch systems. History has shown that the early flight period is of much higher risk than would be expected later in flight history. Figure 5.4.2-2 displays flight reliability for programs "managed by the U.S. government," those managed by foreign governments, and those managed by commercial providers. It should be noted that all of the U.S. vehicles have in large part been engineered and manufactured in industry, and that the so- called "commercial vehicles" were originally developed and produced under government (Air Force) contracts. Importantly, the latter began as ICBMs and, unlike the "government launch vehicles" category, are not human-related.

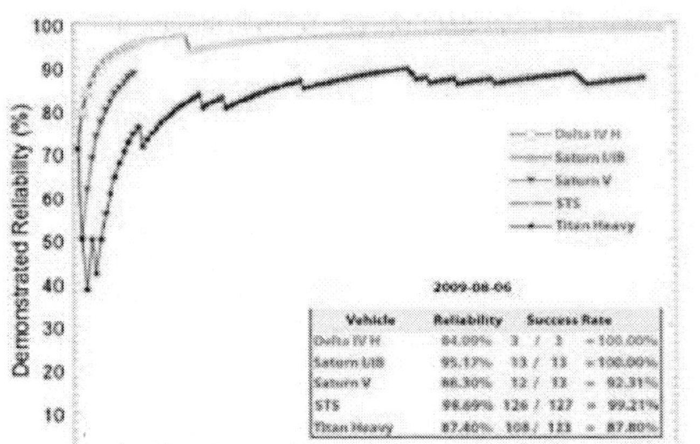

Source: Aerospace Corporation

Figure 5.4.2-1. Historical Growth of Demonstrated Launch Reliability

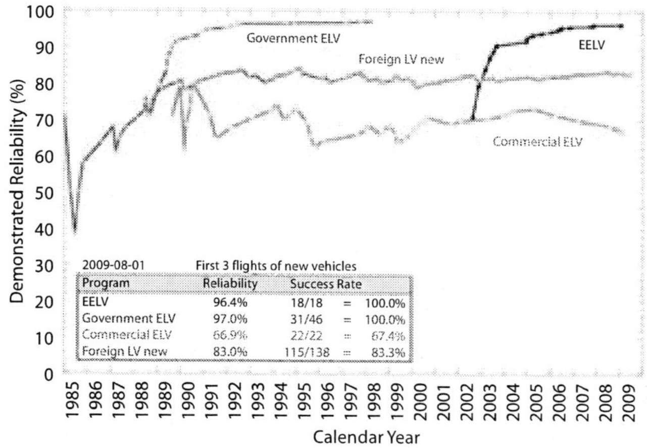

Source: Aerospace Corporation

Figure 5.4.2-2. Demonstrated reliability growth under various programs

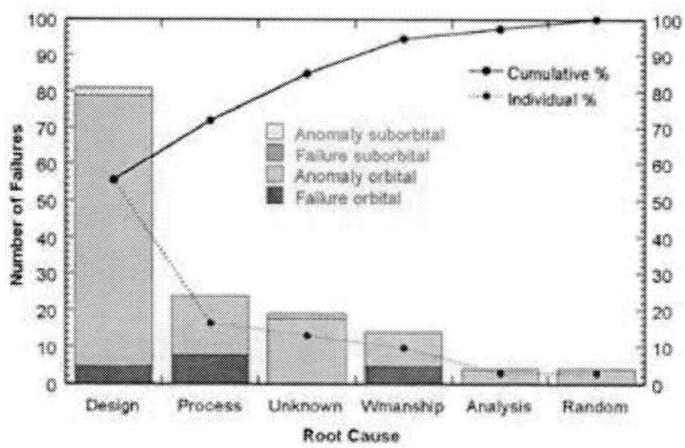

Source: Aerospace Corporation

Figure 5.4.2-3. Root Cause of Launch Vehicle Failures

Space launch vehicles have a history of malfunctions caused by human pre-flight technical error rather than so- called "random" part failure or in-use operator error. This is clearly demonstrated in Figure 5.4.2-3 which shows the root cause of flight malfunctions derived from all U.S. heavy-lift launches to date.

Included in the data are the results of process and design errors associated with the Shuttle Thermal Protection System and SRB joint gas anomalies that were observed— but were not viewed as hazardous prior to the failures of *Challenger* and *Columbia*, respectively. Although some Thermal Protection System anomalies continue to occur, SRB joint gas leaks, which had been a recurring problem with the large segmented solid motors, have been significantly less prevalent after the design changes following the *Challenger* failure.

Summary

There are many issues that must be carefully considered leading to the final decisions on the launch system for both heavy cargo and crew. These include, but are not limited to:

- The cost, schedule and performance of the launch system.
- The likelihood that an increase in volume of production and operation, including by other customers, will decrease costs in the future.
- The impact on the present and future industrial base.
- The initial and ultimate reliability of the launcher, and the extent to which the heritage of the launcher influences the early operational reliability.
- The benefit of independent assurance in increasing the demonstrated reliability.
- The root cause of failures of launchers, and the extent to which these can be modeled and controlled by sound practices in design and processing.

6. Program Options and Evaluation

This chapter presents a synthesized set of options for the future of U.S. human spaceflight. Section 6.1 presents the evaluation criteria the Committee developed. The five key questions that framed the examination are briefly re-stated in Section 6.2, and the choices for each are summarized. These have been presented individually in Chapters 3, 4, and 5. In Section 6.2, pointers denote places where the individual choices will be discussed in the context of the Integrated Options. These analyses form the basis of Sections 6.3 through 6.6. These Integrated Options tie together choices from the five key decisions to allow cost and schedule to be assessed, and an overall evaluation of progress to be made.

6.1. Evaluation Criteria

In order to conduct an independent review of ongoing U.S. human spaceflight plans and alternatives, the Committee recognized that it would be important to define a process that would equitably evaluate the wide range of options to be identified. Consistent with the systems engineering approach, it was important to clearly define the set of criteria against which all options would be assessed, and to define an evaluation process that would enable a fair and consistent assessment of each option. Since many of the evaluation criteria are not quantitative, the Committee did not intend that the evaluation would generate a single numerical score; rather, it would provide a basis for comparison across options, highlighting the opportunities and challenges associated with each. Assigning weights to individual figures of merit is within the purview of the ultimate decision-makers.

In order to identify Integrated Options that are safe, innovative, useful, affordable and sustainable, the Committee developed a number of evaluation criteria by which the relative merits of various human spaceflight missions and objectives could be compared. The Committee was chartered by the Office of Science and Technology Policy in the Executive Office of the President. The Committee's Statement of Task (see Appendix C) provided important guidance. The Committee considered metrics suggested by its members, as well as those based on previous reviews and studies, such as the 1991 Synthesis Group, as well as those derived from policy and historical documents including the Space Act of 1958 and the 2004 Vision for Space Exploration.

There are numerous challenges in evaluating the complex set of Integrated Options considered for human spaceflight, because they vary widely with respect to three principal "dimensions":

- **Benefits to Stakeholders.** The community of stakeholders is diverse, and the potential benefits are equally wide-ranging. Stakeholders include: the U.S. government; the American public; the scientific and education communities; the industrial base and commercial business interests; and human civilization as a whole. Each option offers benefits to some subset of stakeholders, nations and humankind. These benefits include: the capability for exploration; the opportunity for technology innovation; the opportunity to increase scientific knowledge; the opportunity to expand U.S. prosperity and economic competitiveness; the opportunity to enhance global partnership; and the potential to increase the engagement of the public in human spaceflight.

- **Risk.** Each benefit has associated risks. There is uncertainty about the level of benefit that will actually be achieved, since attaining some of the goals may take decades. These risks are not independent of each other. NASA can mitigate some, but others are driven by external forces. Good program management can ameliorate schedule and programmatic risk, given sufficient schedule margin and financial reserves. Safety and mission risk can be managed by changing the mission profile or reducing program content. Programmatic sustainability is a key concern, since any change to the human spaceflight program is likely to affect existing contractual agreements and will require long-term commitments beyond the term of any one presidential administration. There is also risk to the nation's workforce and capabilities in critical skills. Currently, NASA's workforce represents expertise and experience that have enabled its outstanding achievements in space. The industry workforce is fragile, because once the need for a capability stops, marketplace pressures will diminish it, and reconstituting it may be very difficult and extremely costly.

- **Budget Realities.** The desire to identify options that will fit within the existing budget is a significant constraint. Ultimately, only the President and Congress can determine what is affordable in the context of other major national financial demands. Human spaceflight is not a short-term commitment, and it requires budget stability for decades in order to achieve its goals. Year-to-year funding will affect NASA's ability to successfully implement strategic decisions that can reduce total life-cycle cost. The Committee recognizes that operating costs have a sustained and significant impact on the budget, and they may limit the ability to start new efforts critical to achieving human spaceflight goals. For the purpose of this assessment, the Committee chose to represent the full cost of the programs and did not assume any financial contributions by international partners.

These dimensions were expanded into 12 criteria by which Integrated Options could be compared. The Committee clearly recognized that for each option there is some degree of uncertainty as to the magnitude of the influence and interdependence across the three dimensions. The Committee selected the following criteria as the basis for evaluation:

1) **Exploration Preparation.** Since the nature of exploration is, by definition, uncertain and subject to the surprise of discovery, it is important to establish a robust program that provides the opportunity to demonstrate technology, systems and operations that will be important in future exploration—specifically, a program that can be adapted to explore destinations that facilitate missions to Mars.

2) **Technology Innovation.** Integrated Options should enable technology maturation and foster the development of new modes of exploration, in addition to creating new technologies and new engineering knowledge that will enhance exploration. Technology development should also provide the opportunity to demonstrate national leadership in innovation. Technologies that are of use to stakeholders beyond NASA can be critical to the nation as a whole and should be sought during the exploration effort.

3) **Science Knowledge.** Integrated Options should address research areas critical to the scientific community as defined by the National Academies' decadal survey priorities. They should include an implementation plan that supports accomplishment of the prescribed research as a key product of the mission. The Committee recognizes that for some decadal survey priorities, the requirement for the human spaceflight program is simply to do no harm. However, science can be enhanced by human exploration, particularly of complex environments, and by providing the ability to service scientific facilities in space.

4) **Expanding and Protecting Human Civilization.** Integrated Options should lead to the possibility of a sustained off-planet human presence. They should also support research for physiological effects associated with radiation and zero- or low-g, as well as psychological stress associated with long-duration remote exploration. Finally, an option is more highly rated if it will aid in the protection of human civilization against a near-Earth-object impact.

5) **Economic Expansion.** Integrated Options should encourage and stimulate a growing, profitable industrial base. They should provide an opportunity for a sustained commercial engagement, and they should help increase U.S. development and production capabilities. Those capabilities would, in turn, increase the nation's international competitiveness, as well as ultimately lower the cost of space transportation and operation.

6) **Global Partnerships.** Integrated Options should provide the opportunity to strengthen and expand international partnerships in the human spaceflight program. These would include existing international partners, but should not preclude expansion to new partners, and would allow partners to make contributions that

could be on the critical path to mission success. Participation by other countries would be advantageous not only from the perspective of encouraging global cooperation, but also in terms of creating opportunities for synergistic research, risk reduction, cost-sharing and technology interchange.

7) **Public Engagement.** Integrated Options should inspire current and future generations, educate the public about the opportunities and societal benefits gained from space missions, and motivate young people to pursue an education in science, technology, engineering and mathematics, followed by careers that capitalize on this education. Options that provide the opportunity for regular visible accomplishments can galvanize broad public interest in exploration.

8) **Schedule and Programmatic Risk.** Integrated Options should be formulated to deliver a stated exploration capability on schedule. The technical design should be robust, and technologies required should be reasonably mature, with sufficient schedule and technical margins to support risk mitigation.

9) **Mission Safety Challenges.** The Committee did not carry forward any Integrated Option that failed to provide for reasonable crew safety and overall mission success. Therefore, to discriminate among options, the Committee assessed the relative risk and complexity of mission scenarios and the likely impact on crew safety and mission reliability. Human exploration beyond low-Earth orbit will be riskier than current human spaceflight activities, since such exploration requires doing things that have not been done before, and the options for recovery will be limited. Missions that involve beyond- lunar landings, beyond-near-Earth-orbit fly-bys, or complex orbital operations will have mission profiles with a significantly increased safety challenge.

10) **Workforce Impact.** The Committee evaluated the impact on the workforce in two ways: 1) the impact on national critical skills; and 2) the impact on the total NASA and industrial workforce. Each option was assessed for its potential impact on the critical skills across the nation and the ability to retain or develop the needed critical skills and expertise needed both in industry and at NASA. The impact on the total workforce is an assessment of the potential for a reduction of employment as a function of the dollar investment in human spaceflight.

11) **Programmatic Sustainability.** Integrated Options should have a broad base of support for ongoing future funding. They should have a manageable impact on pre-existing contracts and enable a smooth transition from current human spaceflight operations. Support and advocacy for the option could come from other government agencies, space- related organizations, industry and Congress.

12) **Life-Cycle Cost.** The FY 2010 budget for NASA's future human exploration programs, including operations cost through 2020, is roughly $99 billion. The Committee was tasked to provide at least two options within this budget profile, and if appropriate, provide options that are somewhat less-constrained. Clearly, a

program that meets the current budget with high value in other categories would be the ideal in complying with the Committee's charter.

The Committee deliberated at length in public meetings about the advantages and disadvantages of each option with respect to these 12 criteria. Wherever possible, quantitative analytical assessments were utilized to inform the ratings. In the end, however, it was usually necessary to interpret the available information through the considered judgments of the 10 members of the Committee, based on their experience in space matters. The results were captured using a score of -2 (least benefit),-1, 0, 1 or 2 (most benefit) for each attribute of each Integrated Option.

6.2. Key Decisions and Integrated Option

6.2.1. Key decisions

The future of U.S. human spaceflight in the upcoming decades can be formulated in terms of five key questions, and the associated choices for answering each one. The questions, outlined in Chapter 1 and discussed further, with potential answers, in Chapters 3, 4 and 5, are:

1. **What should be the future of the Space Shuttle?**
   Prudent fly-out of remaining flights (currently part of NASA policy, but FY 2011 funding is not part of the President's budget).
   Extend Shuttle through 2015 at minimum flight rate (only likely in conjunction with extending the ISS, and developing a directly Shuttle-derived heavy-lift vehicle).

This question will be examined in Section 6.4, in which some Integrated Options extend the Shuttle life and others do not.

2. **What should be the future of the International Space Station (ISS)?**
   Terminate U.S. participation in the ISS by the end of 2015.
   Continue U.S. participation, through at least 2020 (probably at an enhanced level of U.S. utilization).

This decision will be discussed in Section 6.4, in which some Integrated Options extend the utilization of the ISS by the U.S. and others do not.

3. **On what should the next heavy-lift launch vehicle be based?**
   Ares I plus Ares V.
   Ares V Lite dual launch, with no refueling required for lunar missions, but enhanced with (potentially commercial) refueling for more demanding missions.
   Directly Shuttle-derived vehicle, enhanced with (potentially commercial) refueling.
   Evolved Expendable Launch Vehicle (EELV)-heritage "super heavy" vehicle enhanced with (potentially commercial) refueling.

This discussion is contained in two sections of the chapter. Section 6.4 includes a comparison of the Ares I plus Ares V architecture with the Ares V Lite dual launch architecture. Later, the Ares family and other options are contrasted in Section 6.5.

4. **How should crews be carried to low-Earth orbit?**
   U.S.-government-provided systems.
   Commercially provided systems (with backup by U.S. government system).

Section 6.4 contains Integrated Options that include both choices.

5. **What is the most practicable strategy for exploration beyond low-Earth orbit?**
   Moon First on the Way to Mars, with lunar surface exploration focused on developing capability for Mars.
   Flexible Path to Mars via the inner solar system objects and locations, with no immediate plan for surface exploration, then followed by exploration of the lunar and/or Martian surface.

Section 6.4 contains Integrated Options that have the Moon First as their strategy, while Section 6.5 presents those that explore along the Flexible Path. The cross-case analysis of these options will be presented in Section 6.6.

The Committee addressed one other underlying question: what meaningful exploration program beyond low-Earth orbit could be executed within the budget constraints represented in the FY 2010 budget. This will be the topic of Section 6.3.

The Committee considers the framing and answering of these questions, individually and consistently, to be its principal evaluation of the potential U.S. human spaceflight plans. The Integrated Options were prepared in order to understand the interactions of the decisions, particularly with regard to cost and schedule. By formulating the Integrated Options, the Committee did not mean to constrain the possible final decision, but only to inform it. Other reasonable and consistent combinations of the choices are obviously possible (each with its own cost and schedule implications), and these could also be considered as alternatives. The Integrated Options evaluated are intended to represent the families of options, yet without presenting an unmanageable number of alternatives. The Committee, in keeping with its charter, expresses no preference among these families, but does discuss the various advantages and disadvantages with respect to the evaluation criteria (without weighing those attributes).

6.2.2. Integrated options

The Committee has defined five principal Integrated Options for human spaceflight. These have been selected from the more than 3,000 possible alternatives. Even after defining the choices for the five key decisions, 62 different options are possible. Not all of these combinations are worth considering—for example, it makes little sense to extend the Shuttle life if the heavy lift will be based on EELV heritage.

	Budget	Shuttle Life	ISS Life	Heavy Launch	Crew to LEO
Constrained Options					
Option 1: Program of Record (constrained)	FY10 Budget	2011	2015	Ares V	Ares I + Orion
Option 2: ISS + Lunar (constrained)	FY10 Budget	2011	2020	Ares V Lite	Commercial
Moon First Options					
Option 3: Baseline - Program of Record	Less constrained	2011	2015	Ares V	Ares I + Orion
Option 4A: Moon First - Ares Lite	Less constrained	2011	2020	Ares V Lite	Commercial
Option 4B: Moon First - Extend Shuttle	Less constrained	2015	2020	Directly Shuttle Derived + refueling	Commercial
Flexible Path Options					
Option 5A: Flexible Path - Ares Lite	Less constrained	2011	2020	Ares V Lite	Commercial
Option 5B:Flexible Path - EELV Heritage	Less constrained	2011	2020	75mt EELV + refueling	Commercial
Option 5C: Flexible Path - Shuttle Derived	Less constrained	2011	2020	Directly Shuttle Derived + refueling	Commercial

Source: Review of U.S. Human Spaceflight Plans Committee

Note: Program-of-Record-derived options (Options 1 and 3) do not contain a technology program; all others do.

Figure 6.2.2-1. A summary of the Integrated Options evaluated by the Committee

	Elements 2010-2020	Elements 2021-2030
Constrained Options		
Option 1: Program of Record (constrained)	Shuttle, ISS, Ares I, Orion	Ares I, Ares V, Orion
Option 2: ISS + Lunar (constrained)	Shuttle, ISS, Commercial Crew Transport	Ares V Lite, Orion
Moon First Options		
Option 3: Baseline - Program of Record	Shuttle, ISS, Ares I, Orion	Ares I, Ares V, Orion, Altair, Lunar Surface Systems
Option 4A: Moon First - Ares V Lite	Shuttle, ISS, Commercial Crew Transport	Ares V Lite, Orion, Altair, Lunar Surface Systems
Option 4B: Moon First - Extend Shuttle	Shuttle, ISS, Commercial Crew Transport	Directly Shuttle Derived Launcher, Orion, Commercial Crew Transport, On-Orbit Refueling, Altair, Lunar Surface Systems
Flexible Path Options		
Option 5A: Flexible Path - Ares V Lite	Shuttle, ISS, Commercial Crew Transport	Ares V Lite, Orion, On-Orbit Refueling, In-space Habitat and Propulsion, Hybrid Light Lunar Lander, Light Lunar Surface Systems
Option 5B: Flexible Path - EELV Derived	Shuttle, ISS, Commercial Crew Transport	EELV Derived Super Heavy, Orion, Commercial Crew Transport, On-Orbit Refueling, In-space Habitat and Propulsion, Hybrid Light Lunar Lander, Light Lunar Surface Systems
Option 5C: Flexible Path - Shuttle Derived	Shuttle, ISS, Commercial Crew Transport	Directly Shuttle Derived Launcher, Orion, Commercial Crew Transport, On-Orbit Refueling, In-space Habitat and Propulsion, Hybrid Light Lunar Lander, Light Lunar Surface Systems

Source: Review of U.S. Human Spaceflight Plans Committee

Figure 6.2.2-2. Principal elements developed or used in the Integrated Options

The five Integrated Options considered by the Committee are: one Baseline case, founded on the current Constellation Program, plus four alternate options, summarized for reference in Figure 6.2.2-1. The first two Integrated Options represent attempts by the Committee to develop alternatives that are compatible with the FY 2010 budget profile. Each follows an approach that is principally aimed at lunar exploration. Option 1 uses the content of the Program of Record, while Option 2 extends the ISS, uses commercial crew delivery, skips development of the Ares I, and uses the Ares V Lite as the sole NASA launcher. (See Figure 6.2.2-2.) The Baseline (Option 3) and the remaining two options are all fit to the same, less-constrained budget profile. Options 3 and 4 are also lunar-orientated strategies. Option 3 is an implementable version of the Program of Record, with nearly exactly the content of the

Constellation Program, but with two minor changes in funding that the Committee found necessary: to extend the Shuttle into FY 2011, and to deorbit the ISS in 2016. Option 4 again extends the ISS. Lite and commercial crew. Option 5 is the one based on the Flexible Path. Integrated Options 4 and 5 are explored in several variants that principally examine the sensitivity to the heavy-launch systems. A more visual representation, showing a possible decision logic among these options, is shown in Figure 6.2.2-3.

6.2.3. Methodology for analyzing the integrated options

The committee used a synthesized cost-schedule-value methodology to assess the Integrated Options. In the first step, the Committee created the Integrated Options that have been presented in Figure 6.2.2-1. These options were intended to link together representative combinations of the outcomes. The Committee included the Baseline case of the Program of Record in the mix. The other Integrated Options were chosen to allow some of the key trades (destination, launch vehicles, etc.) to be contrasted.

The Committee used two candidate budget profiles for examining the Integrated Options. In the first, the guidance of the FY 2010 budget was enforced. This is called the "Constrained Case" or simply FY 2010 Budget Case, as shown in Figure 6.2.3-1. It became apparent to the Committee that options were needed that were not constrained to the FY 2010 budget. For planning and evaluation purposes, the Committee created a second budget profile that rose from the FY 2010 budget number to a sum \$3 billion higher in 2014, and then rose at an expected inflation rate of 2.4 percent thereafter (Figure 6.2.3-1). Thus, by combining the five key decisions and the two budgetary scenarios, the Committee produced the five Integrated Options, with variants, that are listed in Figure 6.2.2-1.

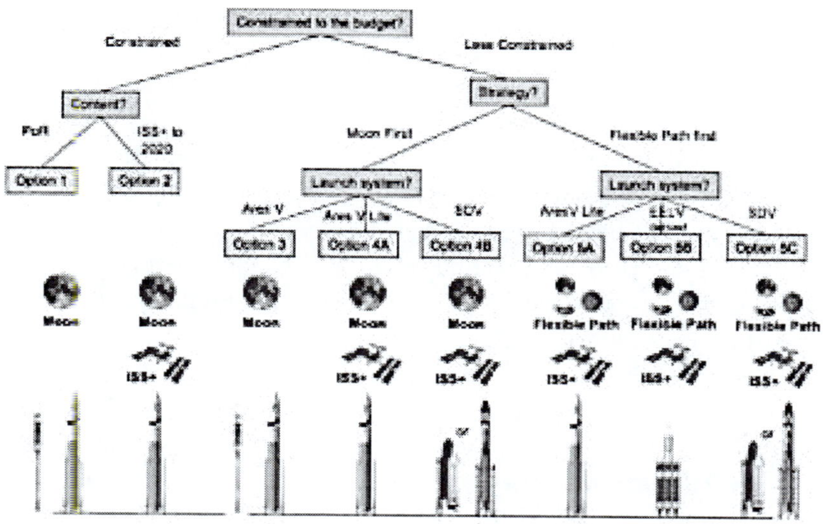

Source: Review of U.S. Human Spaceflight Plans Committee

Figure 6.2.2-3. Suggested Integrated Option decision tree, showing the principal features of the options

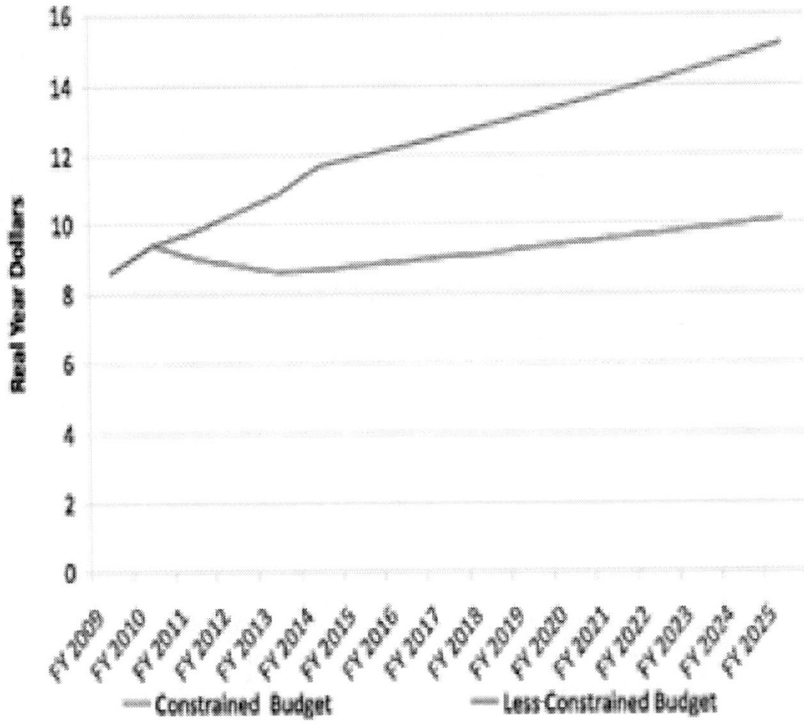

Source : Review of U.S. Human Spaceflight Plans Committee

Figure 6.2.3-1. Two budgetary assumptions used by the Committee: the constrained budget which follows the guidance in the FY 201 0 budget, and a less-constrained planning budget developed by the Committee

To assess the benefits and risks of these Integrated Options, the Committee next applied the evaluation criteria that had been selected (Section 6.1). In order to assess cost and schedule implications, the Committee then examined the Integrated Options using the affordability analysis tool. The Committee created a set of assumptions and ground rules for this analysis. These included budget profiles for the constrained and less-constrained budgets and input manifests for each option.

The costs used by the Committee are intended to show the full cost of a program, and do not take into account any potential contribution by international partners. Much of the cost data input into the analysis originated from the Constellation Program. In those cases where the program cost estimates already contained a credit for an international contribution, this credit was removed by the Committee in order to show full cost. In other cases, the Committee itself or other sources helped to define the input. For example, the EELV super-heavy-launcher costs were based on Aerospace Corporation examination of the EELV development plans. The organizational transition costs in Option 5B were provided by NASA Headquarters. The in-space habitat in Options 5 was the result of an assessment of comparables performed by Aerospace. The Committee provided estimates for the commercial crew-to-low-Earth orbit costs, as discussed in Section 5.3.3.

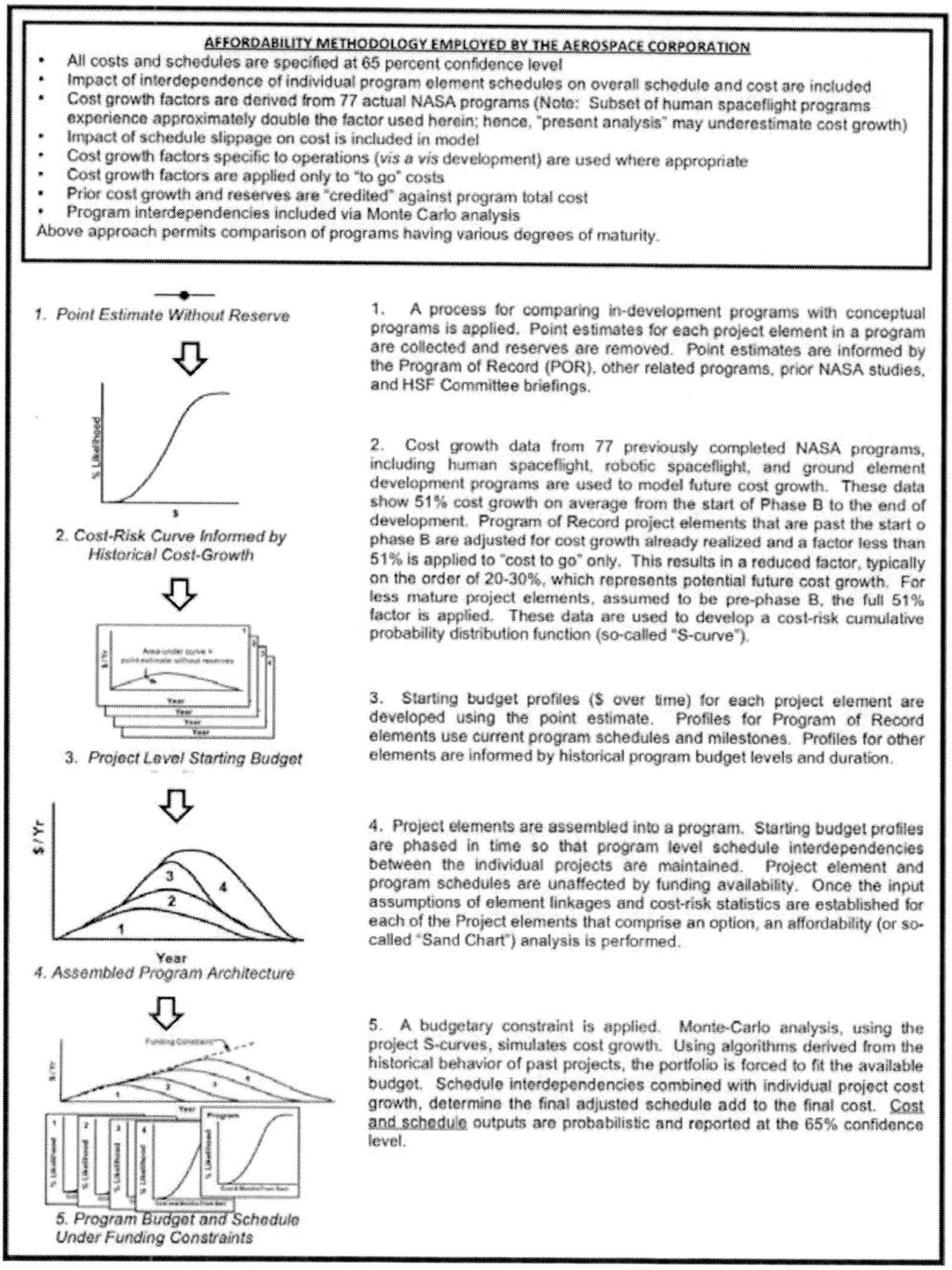

Source: Aerospace Corporation

Figure 6.2.3-2. Cost and Schedule methodology used by the Aerospace Corporation for affordability analysis of Integrated Options

The Aerospace team conducted the affordability analysis using the process described in Figure 6.2.3-2. This analysis outputs key dates and element costs at the 65 percent confidence level. It also estimates the uncertainty on dates and costs. Output manifests are derived based on the 65th percentile confidence level, and are illustrative of the pace of missions and elements utilized. The affordability analysis corrects the input cost in several ways. First, it estimates a range of expected growth of the cost for each program element from System Design Review (SDR, Start of Phase B) to completion, based on historical data of NASA programs. At the average, this introduces a 51 percent growth from the estimate held at SDR in the cost for development (DDT&E costs). For elements that have not reached their SDR, such as the Ares V or commercial crew service, this full correction was applied. For elements that have passed their SDR, credit was given for subsequent development and maturity of the design. For example, the mean cost of the Orion in the analysis, due to this factor, is only 25 percent higher than would be reported by the Program of Record *at the mean*. Other, more mature programs, such as the Ares I, receive credit by a similar process. In operations, a 26 percent growth factor was applied to unproven systems, and no growth factor at the mean was applied to existing systems such as the Shuttle or the ISS, or to defined budget items such as the technology program.

NASA Headquarters asked the Program of Record to report cost and schedule at the 65 percent level, and the Committee attempted to report in a consistent manner. Note that on average, the difference between the mean of expected costs and the 65 percent confidence costs adds about 10 percent to all program costs calculated. Finally, the affordability analysis combines the development schedule of all the elements of the program. This process accounts for the additional cost to one element if another element it depends upon slips in its schedule. This integration of elements typically adds about an additional 10 percent to the total program costs, higher in more-constrained budgets, and lower in less-constrained budgets.

The Committee then examined the outputs of the affordability analyses, and it made interpretations to extract from them the primary information of interest, recognizing the inherent uncertainty in the analysis. The reporting by the Committee attempts to focus on its interpretation of the key milestones and associated uncertainties, and the pace of events after the initial milestone.

6.2.4. Reference cases of the entirely unconstrained program of record

Unconstrained program of record

As an example of the affordability methodology applied to an actual integrated scenario, the reference case of an "implementable" version of the current Constellation Program of Record, unconstrained by *any* budget whatsoever, was analyzed. This implementable version contains only two slight variations from the actual program, instituted by the Committee: the provision for the Shuttle to be flown out in 2011 and additional funds for the deorbit of the ISS in 2016, after withdrawal of U.S. participation at the end of 2015. Note that the ISS is not extended to 2020 in this particular reference case.

As assessed by the Committee, this case delivers Ares I/ Orion in late 2016, achieves human lunar return by the early 2020s, and a human-tended lunar outpost a few years later. These are very close to the dates held internally by the Constellation Program. However, the Committee's analysis indicates that in order to achieve the milestones on that schedule, the implementable Program of Record requires, in real-year dollars (stated at 65 percent confidence):

- About $145 billion over the period from 2010 to 2020, which is:

- About $45 billion over the guidance of the President's FY 2010 budget through 2020, and
- About $17 billion more than what is provided in the "less-constrained budget."
- The expenditures reach $14 billion per year in FY 2016, about $2 billion above the "less-constrained budget" and $5 billion over the FY 2010 budget for that year.
- The expenditures reach over $16 billion per year at their peak in FY 2019, $3 billion above the "less- constrained budget" and $7 billion over the FY 2010 budget for that year.

Thus, both the Program of Record, as assessed by the Constellation Program, and the unconstrained implementable version of the Program of Record, as assessed by the Committee, deliver Ares I and Orion in the mid-to-late 2010s, and they both have human lunar return in the early

2020s. Neither provides for extension of the ISS, or a space technology program of significance. The Committee's finding is that the totally unconstrained implementable version of the Program of Record would significantly exceed even the "less-constrained budget."

Unconstrained program of record with the ISS extension

This case would be identical to the above version of the Program of Record, but with the extension of the ISS to 2020. Since the budget for this case is unconstrained, no milestones slip, but more funds must be added to the NASA budget to operate the ISS in the years between 2016 and 2020.

In this reference case, the Ares I/Orion again are delivered in late 2016, human lunar return is accomplished by the early 2020s, and a human-tended lunar outpost is developed a few years later, again close to the dates held internally by the Constellation Program. The ISS is extended to 2020. But in order to achieve these milestones, this variant of the Program of Record requires, at a 65 percent confidence level, and in real-year dollars:

- About $159 billion over the period from 2010 to 2020, which is:
- About $59 billion over the guidance of the President's FY 2010 budget through 2020, and
- About $31 billion more than provided in the "less- constrained budget."
- The expenditures reach $15 billion per year in FY 2016, about $3 billion above the "less-constrained budget" and $6 billion over the FY 2010 budget for that year.
- The expenditures reach about $19 billion per year at their peak in FY 2019, $6 billion above the "less- constrained budget" and $10 billion over the FY 2010 budget for that year.

Furthermore, the technology budget is a small fraction of a billion dollars each year.

Although these two reference options represent the greatest continuity from the existing Program of Record, the Committee did not include them in the Integrated Options because they greatly exceed the FY 2010 budget profile and because the Committee does not consider them to be programmatically competitive with the Integrated Options discussed below.

6.3. Integrated Options Constrained to the Fy 2010 Budget

6.3.1. Evaluation of integrated options 1 and 2

The Committee was asked to provide two options that fit within the FY 2010 budget profile. That funding profile is shown in Figure 6.2.3-1 It is essentially flat or decreasing through 2014, then increases at 1.4 percent per year

Option 1. Program of record as assessed by the committee, constrained to the FY 2010 budget

This option is the Constellation Program of Record, with only three changes: providing funds for the Shuttle into FY 2011; including sufficient funds to de-orbit the ISS in 2016, and constraining the expenditures to the FY 2010 budget. Under this option the Shuttle retires in FY 2011, and until its retirement in 2015, international crew carriers are used to rotate U.S. crews to the ISS. When constrained to the FY 2010 budget profile, Ares I and Orion are not available until the latter years of the 2010s, after the ISS has been de-orbited, as shown in Figure 6.3.1-1. Starting in the late 2010s, piloted flights in the Ares I and Orion could begin at a pace of several flights per year, but with no specific destination defined. The heavy-lift Ares V is not available until the late 2020s, allowing only orbital flights to the Moon. In addition, there are insufficient funds to develop the lunar lander and lunar surface systems until well into the 2030s, if ever. (See Figure 6.2.2-2.) In short, this program operates within current FY 2010 budget constraints, but offers little or no apparent value.

Option 2. The ISS and Lunar Exploration, constrained to FY 2010 budget

This option extends the ISS to 2020 and begins a program of lunar exploration using Ares V Lite in the dual launch mode. The option assumes Shuttle fly-out in FY 2011, and it includes a technology development program, a program to develop commercial crew services to low- Earth orbit, and funds for enhanced utilization of the ISS. As shown in Figure 6.3.1-1, the Shuttle retires in FY 2011, and international providers rotate crew to the ISS until U.S. commercial crew services become available in the mid-to-late 2010s. Those providers are used to rotate the ISS crew until the Space Station's retirement in 2020. This option does not deliver heavy-lift capability with the Ares V Lite plus Orion until the late 2020s and does not have funds to develop the systems needed to land on or explore the Moon. (See Figure 6.2.2-2.)

The Committee applied its evaluation criteria to assess Integrated Options 1 and 2, as shown in Figure 6.3.1-2. The analysis shows that Option 2, the ISS and Lunar Exploration, outperforms or equals Option 1 in all criteria. It has an equal (but low) Exploration Preparation rating, and performs equally on Science Knowledge, Public Engagement, Schedule Performance, Mission Safety Challenge, and Workforce Impact criteria. Option 2 is more highly rated under several criteria:

- The technology investment provides a higher rating in Technology.
- The extension of the ISS improves its ratings in Human Civilization because of the added micro-gravity human physiology experience gained aboard the ISS, and in Global Partnerships because of the continued engagement of the international partners.

- Use of the commercial crew produces higher scores for Economic Development, and an improved score due to lower Life Cycle Costs.
- The resulting stronger advocacy increases the rating for Sustainabiity.

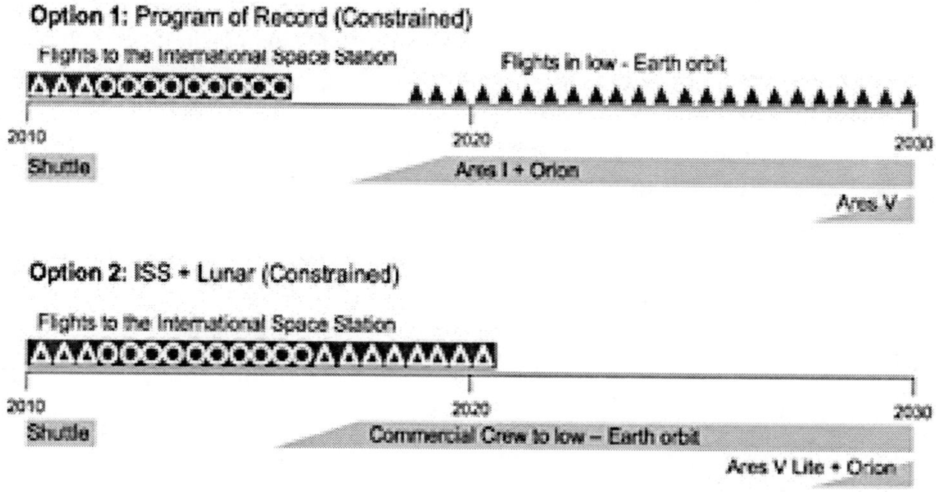

Source: Review of U.S. Human Spaceflight Plans Committee thereafter, which is slower than the 2.4 percent inflation rate used by the Committee.

Figure 6.3.1-1. Timelines of the two Integrated Options constrained to the FY 2010 budget-milestones (above the line) indicate pace of activity, while ramped bars (below the line) indicate approximate range of uncertainty in the date at which the capability becomes available

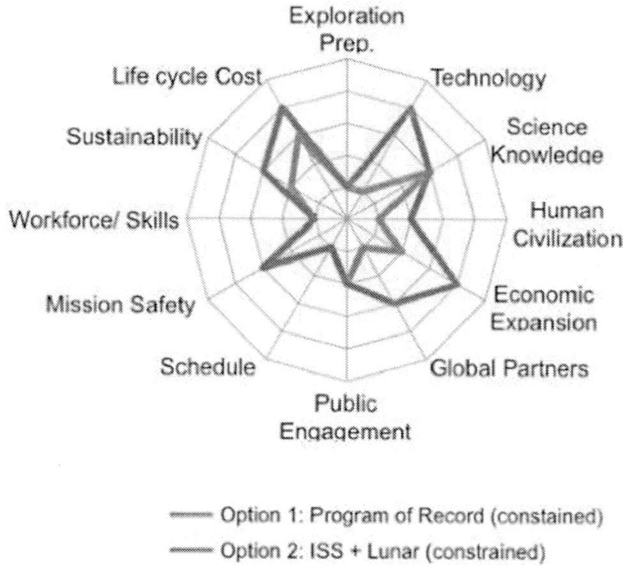

Source: Review of U.S. Human Spaceflight Plans Committee

Figure 6.3.1-2. Relative evaluation of factors under Option 1 (implementable Program of Record) and Option 2 (ISS + Lunar), both constrained to the FY 2010 budget profile. Options closer to the center are of lesser value

6.3.2. Examination of alternate budget guidance

The first two Integrated Options allow examination of the underlying question: what meaningful exploration program beyond low-Earth orbit could be executed within the budget constraints represented in the FY 2010 budget? The Committee concludes that two executable alternatives comply with FY 2010 funding guidance. Option 2 is scored on the evaluation criteria more highly than Option 1, which provides for neither the ISS extension nor technology development, but neither alternative provides for a viable exploration program.

In the process of developing these options, the Committee conducted a sensitivity analysis to determine whether any reasonable exploration program (e.g., with different heavy-lift vehicles, or a different exploration destination) would fit within the FY 2010 budget guidance. The Committee could find none. In addition, the Committee tried to develop a variant of the Flexible Path that fit within the FY 2010 budget, and such a variant looked no more promising than Option 2, with the first missions beyond low-Earth orbit in the late 2020s.

This analysis led the Committee to its finding that human exploration beyond low-Earth orbit is not viable under the FY 2010 budget guideline. It would be possible to continue the ISS and a program of human activity in low-Earth orbit within this budget guidance, and to develop the technology for future exploration, but the budget limitation would delay meaningful exploration well into the 2020s or beyond.

6.4. Moon First Integrated Options Fit to the Less-Constrained Budget

6.4.1. Evaluation of integrated options 3 and 4

Option 3 and Option 4 (and its variants) pursue the Moon First exploration strategy, but are not constrained to the FY 2010 budget profile. Rather, they are fit to the "less-constrained" planning budget that the Committee developed. That budget profile, shown in Figure 6.2.3-1, increases to $3 billion above the FY 2010 guidance between FY 2011 and FY 2014, and then grows with inflation at an expected inflation rate of 2.4 percent per year.

While it was formulating Integrated Options, the Committee quickly realized that viable options could not be found within the constrained budget. It then examined potential increases in the budget that would enable a sustainable and executable human spaceflight program. By examining several different potential expenditure profiles, the Committee arrived at the above investment level that would provide for the extension of the ISS, allow progress towards exploration beyond low-Earth orbit, and make an investment in technology. It provided a useful standard by which various options could be compared in a meaningful way.

Option 3. Baseline Case: Implementable program of record

This is an executable version of the Program of Record. (See Figure 6.2.2-2.) It consists of the content and sequence of that program: de-orbiting the ISS in 2016; developing Orion, Ares I and Ares V; and beginning exploration of the Moon. The Committee made only two additions it felt essential: budgeting for the fly-out of the Shuttle in 2011 and including additional funds for the ISS de-orbit. The Committee then applied the less-constrained budget profile.

The Committee's assessment of the schedule outcome for this option is shown in Figure 6.4.1-1. Under that schedule, the Shuttle retires in FY 2011, and international crew services ferry U.S. crews to the ISS until its retirement in 2015. The option delivers Ares I/Orion in the mid–to-late 2010s, and flights to low-Earth orbit, but with no specific destination as yet definable. The Ares V becomes available, and human lunar return occurs in the mid-2020s. With a pace of about two flights per year, a lunar base begins to function about three years later.

Although not included by the Committee as an Integrated Option, a variant of Option 3 is possible that extends the ISS to 2020, adds the technology program, and maintains all of the other content of Option 3. Some would argue that this variant is actually the reference program on which NASA is embarked. But it should be emphasized that so far, no funds have specifically been allocated for continuing the ISS after 2015. The Constellation Program, in its planning, assumed that funds that had been previously used for the operation of the ISS would become available to Constellation by 2016. Additionally, no funds were explicitly in the NASA plans for a robust, broad-based technology development program.

The impact of extending the ISS to 2020 is to require the additional expenditure of about $14 billion between 2015 and 2020. This additional expenditure, plus the technology program, combined with developing the Ares I and Orion within the "less-constrained budget," causes significant slips to subsequent milestones. Orion and Ares I do not become available until the late 2010s, serving the last few years of the ISS. The development of heavy lift and its use in human lunar return slips to the late 2020s, so that the Ares I and Orion are left to either not fly at all for the better part of a decade, or fly in that interval in Earth orbit without the ISS as a destination. The Committee observed that the other options would be rated more highly by the evaluation process, so the Committee did not pursue this variant of Option 3.

Option 4. Moon first

This option preserves the Moon as the first destination for human exploration beyond low-Earth orbit. It also extends the ISS to 2020, funds technology advancement, uses commercial vehicles to carry crew to low-Earth orbit and funds the Space Shuttle into FY20 11. There are two significantly different variants to this option. (See Figure 6.2.2-2.)

Variant 4A is the Ares V Lite variant. As shown in Figure 6.4.1-1, this option retires the Shuttle in FY 2011 and relies on international launch support for crew delivery until the U.S. commercial crew services become available in the midto-late 2010s. The commercial crew provider ferries crew to the ISS until its retirement in 2020, after which there is a gap in human flight activity until the Ares V Lite in the dual launch mode is available for lunar exploration in the mid- 2020s. The beginnings of a lunar base follow about three or four years later.

Variant 4B is the Shuttle extension variant. This variant includes the only foreseeable way to close the gap in U.S. human-launch capability. As shown in Figure 6.4.1-1, this variant extends the Shuttle to 2015 at a minimum safe-flight rate. Shortly after the Shuttle is retired, commercial service picks up the ferrying of crew to the ISS until its retirement in 2020. This variant also takes advantage of synergy with the Shuttle by developing the more directly Shuttle-derived heavy-lift vehicle. However, even with the "less-constrained budget," the more directly Shuttle-derived heavy launcher is not available until the middle of the 2020s, 10 years after the Shuttle retires. Therefore, the problems of continuity of systems and workforce skills are diminished in the early years, but reappear later. The more directly Shuttle-derived system has a higher recurring cost, so the flight rate within the "less-constrained budget" drops below two crew flights per year to the Moon, and a lunar base slips beyond 2030.

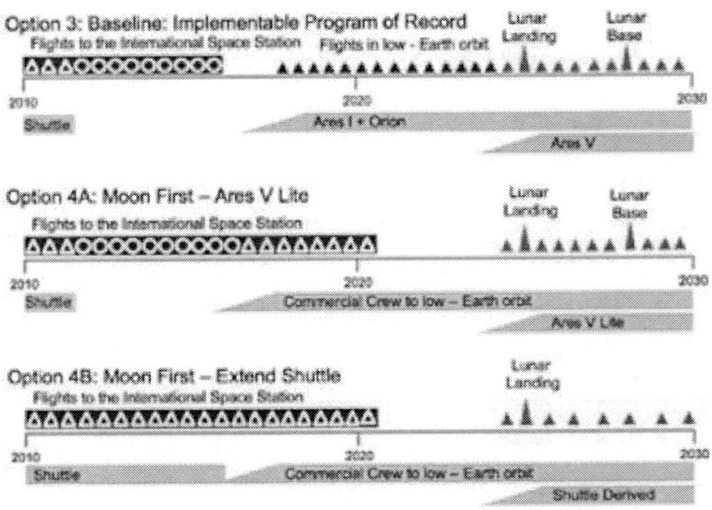

Source: Review of U.S. Human Spaceflight Plans Committee

Figure 6.4.1-1. Timeline of the two Integrated Options (and variant) that follow the Moon First strategy, fit to the less-constrained budget. Milestones (above the line) indicate pace of activity, while ramped bars (below the line) indicate approximate range of uncertainty in the date at which the capability becomes available

Source: Review of U.S. Human Spaceflight Plans Committee

Figure 6.4.1-2. Relative evaluation of factors in Option 3 (Baseline –Program of Record) and Option 4A (Moon First – Ares V Lite), subject to the less-constrained budget. Options closer to the center are of lesser value

Source: Review of U.S. Human Spaceflight Plans Committee

Figure 6.4.1-3. Relative evaluation of factors in Option 4A (Moon First – Ares Lite) and Option 4B (Moon First – Extend Shuttle), subject to the less- constrained budget. Options closer to the center are of lesser value

Again the Committee applied its valuation process to these Integrated Options and variants that fit within the less-constrained budget and follow a Moon First strategy. The options can be compared in two steps. Figure 6.4.1-2 shows the comparison between Option 3, the Baseline, and Option 4A, the Ares V Lite dual launch variant of the Moon First architecture. This shows that Option 4A matches or surpasses Option 3 in all metrics considered. The sources of these differences are:

- The use of the more capable Ares V Lite dual launcher, coupled with the extension of the ISS, providing for more Exploration Preparation.
- The technology investment causes a higher score on the Technology evaluation.
- The extension of the ISS improves the ratings on Human Civilization (more information on human longterm adaptation to space) and Global Partnerships.
- The availability of commercial crew launch gives an edge in Life-Cycle Costs and in Economic Expansion.

Thus the Committee finds that even with the Ares V family of launchers, and the Moon as the destination, there are ways potentially to extract more value from the program than to follow the Baseline.

In a second comparison, Figure 6.4.1-3 shows the relative valuation of the two variants of Option 4, the Moon First. Here the scores indicate some counts on which the Ares V Lite variant 4A scores better than the Shuttle-derived variant 4B:

- The use of the more capable Ares V Lite dual launcher gives a higher score in Exploration Preparation.
- The use of the more economical Ares V Lite gives a better evaluation in Life-Cycle Costs.

- The more capable Ares V Lite dual launch allows simplification of launch and on-orbit operations, reducing Mission-Safety Risk.

In contrast, the more directly Shuttle-derived launcher in Option 4B scores better in Sustainability and Workforce Skills, both traceable to the continuation of the Shuttle system and workforce. This decision trades a more capable vehicle for more short-term benefit from advocacy and smaller workforce impact with the Shuttle-derived systems.

6.4.2. Examination of the key decision on the ISS extension

Comparison of Integrated Options 3 and 4A allow examination of the key decision concerning the future of the International Space Station: Should we stop U.S. participation in the ISS at the end of 2015, or continue U.S. participation, through at least 2020 (probably at an enhanced level of U.S. utilization)? The background for this question was presented in Section 4.2.

Extending the ISS would yield several benefits; chief among these is the support for global partnerships. By ending participation in 2015, the U.S. would voluntarily relinquish its leading role in this phase of international space exploration. By extending the ISS, we would further develop the international partnerships upon which the ISS is based, encouraging these ties to evolve into long-lasting relationships for space exploration. The return on investment on the part of the U.S. would be enhanced by 10 years of well-funded utilization in the 2010s, and by the operation of as a National Laboratory.

The ISS extension issue couples with Exploration Preparation. If properly used, the ISS could be a more effective testbed for development of the technologies and systems for exploration. Extending the ISS also leads to improved evaluation in Human Civilization. The extra years on the ISS would allow a better understanding of the adaptation of humans to micro-gravity, and the extension would produce more data for extremely long stays, important in planning for exploration.

The choice of ending U.S. participation in the ISS in 2105 really provides only one benefit, that of freeing up the roughly $2.5 to $3 billion per year needed to run the ISS, which can then be invested in the more rapid development of the exploration systems. The Committee's Integrated Option analyses show that if coupled to the choice of commercial crew launch system to low-Earth orbit and the Ares V Lite heavy lift choice, this expenditure on the ISS would delay the exploration of the Moon until the mid-2020s, only a few years after the most aggressive, unconstrained profile would accomplish it. (See Section 6.2.4.)

By applying the evaluation criteria it developed, the Committee finds extension of the ISS to 2020 to have greater value than the choice of ending U.S. participation in 2015.

6.4.3 Examination of the key decision on Ares V vs. Ares V Lite dual launch

Unlike the other four key decisions, each of which has been reduced in this chapter to two main choices, the decision on heavy lift is more complex, as indicated in Figure 5.2-1. The decision can be represented by three successive choices (Figure 6.4.3-1):

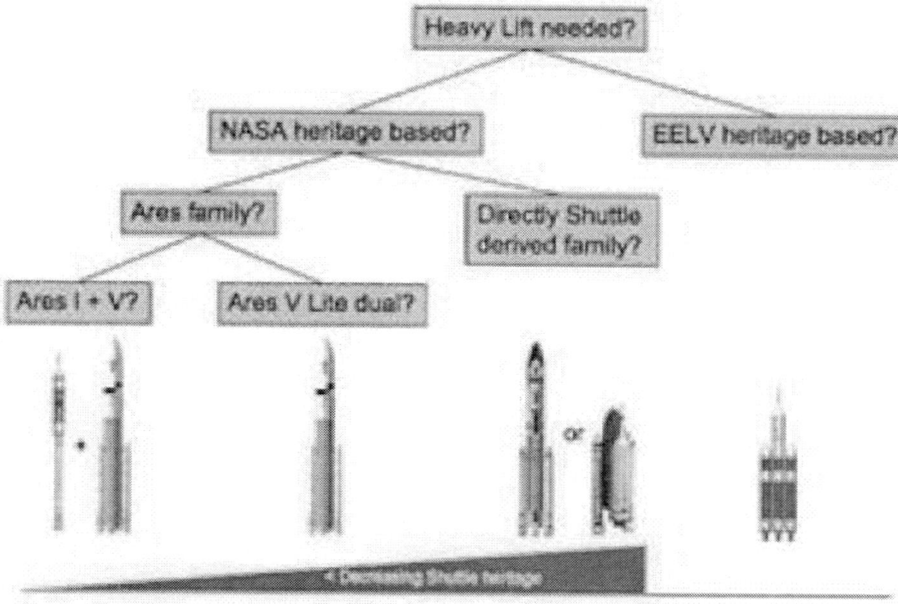

Source: Review of U.S. Human Spaceflight Plans Committee

Figure 6.4.3-1. Suggested launch vehicle decision tree, showing the decisions within the Ares family, between Ares and Shuttle-derived launchers, and between NASA and EELV-heritage launchers

- At the highest level, the nation faces a choice of basing the heavy-lift capability on the NASA-heritage systems (Apollo and Shuttle) or on the EELV-heritage systems.
- Within the NASA-heritage systems are those more directly based on the Shuttle, and those that belong to the Ares family.
- Within the Ares family, there is the choice of Ares I plus Ares V system, currently planned, or the Ares V Lite (used in the dual mode for lunar missions) as the only vehicle developed.

The issues behind this set of decisions were discussed in Section 5.2. In this section, the third choice will be examined. Because of the structure of the Integrated Options, the best head-to-head comparison of the other decisions is in the Flexible Path options in Section 6.5.

The Committee considered the comparison of the Ares V Lite, which for lunar missions would be used in dual mode, with the current architecture of Ares I plus Ares V. Setting aside the issues of crew launch on the Ares I, which will be discussed in Section 6.4.4, the principal distinguishing feature of the two alternatives is the extra-heavy cargo launch capability that would be provided by the Ares V Lite dual launch during lunar exploration. A secondary distinction is the relatively greater capability of the Ares V compared with the relatively lower performance demands placed on the Ares V Lite.

The main criteria involved in this comparison are Exploration Preparation, Schedule and Program Risks. The current Baseline Ares V has more launch capability than the Saturn V, but current NASA studies show that when used in combination with Ares I, it does not have enough launch capability to robustly deliver the currently planned landing and surface systems to the Moon. In order to deliver the greatest potential of the vehicle, many of the options to increase its performance have already been applied, including using 5.5-segment solid-rocket boosters (SRBs) and six RS-68-family engines (both of which require further development). There is a concern, expressed by some within NASA and by the Committee, that with performance already at the upper edge of what any Ares-V-family vehicle is likely to deliver, coupled with the potential weight growth of the payloads, the development will potentially face delays and added costs associated with weight-reduction campaigns.

In contrast, the Ares V Lite backs off on proposed performance by using a five-segment SRB (already in development) and five RS-68-family engines. Thus the developmental risk of the Ares V Lite is somewhat reduced. The essential difference between the two options is the use of two heavy lifters that for a lunar mission will give substantial margin, and likely save development schedule and cost. The use of the two launch vehicles now decouples the operation of the Orion and Altair. The Ares V Lite dual launch allows a more robust Orion to be built, one that is capable of missions on its own—for example, launch with an Earth Departure Stage (EDS) to a near-Earth object (NEO), which will be discussed in Section 6.5.

Programmatically, the choice of the Ares V (together with Ares I) unquestionably has less impact on current workflow or contracts. However, the Ares V Lite preserves some of the investment already made for Ares I, and would possibly allow some of the contract structure to stay in place. It would use the same five-segment SRB as the Ares I and the same J2-X engine for the Earth Departure stage. It would deliver a heavy-launch vehicle three to five years earlier than if the Ares I were built first (assuming the ISS is extended in both cases). Finally, construction of the Ares V Lite focuses NASA on the more challenging task of building the vehicle most needed for exploration, the heavy-lift booster.

The analysis performed for the Committee indicates that the Ares V Lite dual-launch scenario and the Ares I + Ares V scenario would have comparable operating costs. In a normal year of lunar exploration, for example, in the mid- 2020s, there would be four missions to the Moon, two with crew and two with cargo. In the Ares I + Ares V architecture, this would require two Ares I launch vehicles and four Ares Vs. In the Ares V Lite architecture, this same set of missions would require six Ares V Lites. Considering NASA's high fixed recurring costs, the difference in total cost between four and six launches a year of the same system would be a small fraction of the annual cost. In the Ares I + Ares V architecture, NASA would have two of the six launches by a less expensive vehicle, but would have to operate two launch systems, with two processing flows at the Kennedy Space Center—offsetting the potential savings. While the Ares V would nominally not have to be human-rated, the criticality of the payloads it would carry, and the NASA development culture, would likely (and appropriately) drive it to a nearly human-rated status.

Ares I may have higher single-launch ascent safety than Ares V. Both would be high-reliability rockets, with the same capsule and launch escape system. The first stage of the Ares I is considerably simpler; however, because of the higher dynamic pressure in the flight profile of the Ares I, and its solid-rocket motor, a capsule on the Ares I would have a more challenging separation from the booster than a capsule on the Ares V. Ares I is not planned to launch at a rate higher than two per year, raising questions about the sustainability of safe operations. In contrast, up to six Ares V Lites are planned to launch each year, about the average rate of Shuttle launches throughout that program's years of operation, contributing to a potentially higher demonstrated reliability.

There is widespread confusion about the findings of the Columbia Accident Investigation Board (CAIB) on the issue of mixing crew and cargo. The CAIB report said, "When cargo can be carried to the space station or other destinations by an expendable launch vehicle, it should be." (page 211 of the CAIB Report). That suggests that humans should not be put at risk to carry cargo (as they are in the Shuttle). The implication is *not* that humans should not be launched along with cargo if that makes sense (which was the case with Apollo).

Of the two vehicle choices, the Committee finds the Ares V Lite in the dual mode contributes to a higher score on the evaluation of options than does the Ares V. The critical difference is the use of the two Ares-V-family launchers for lunar missions. Even if the Ares I were to be built, the Committee's findings indicate that the exploration missions would benefit from using the Ares V Lite in the dual mode as described.

6.4.4. Examination of the key decision on the provision of crew transport to low-Earth orbit

The key question pertaining to crew launch to low-Earth orbit is whether to carry the crews on systems provided by the U.S. government or on commercially provided systems (with eventual operational backup by a U.S. government system).

The Exploration Systems Architecture Study (ESAS) of 2005 developed a plan to launch crew to the ISS, and to destinations beyond low-Earth orbit, using the Ares I. As discussed in the background on this decision in Section 5.3, this would be a launch system with very high ascent safety, would have high component commonality with the Ares V, and would provide NASA organizationally with an opportunity to develop the Ares I before undertaking the more complex Ares V.

The alternative is to terminate the development of the Ares I and instead proceed with development of a commercial launch service to low-Earth orbit for crew. If based on a high-reliability rocket, and with a capsule and launch escape system, this approach too could have high ascent safety. It would also have the potential for significantly lower development cost, and therefore be available about a year sooner. Once operating, it would have the potential for significantly lower recurring costs, allowing the more rapid development of systems for exploration beyond low- Earth orbit in a constrained resource environment. The development of commercial crew service is not without significant programmatic risk, as discussed in Section 5.3.

The choice of Ares I as the crew launcher was probably a sound choice in 2005. As is often observed, the rocket equation has not changed, so any reason that NASA would come to a different solution for crew transport to low-Earth orbit today than in 2005 would be due to changes in assumptions and constraints. The Committee in fact concludes that many of the assumptions on which the Ares I crew decision was based have changed. In contrast, the Committee found that the Orion should continue to be developed as a capable crew exploration vehicle, regardless of the decision on Ares I. Likewise, it should be emphasized that the Committee did not find any insurmountable technical issues with Ares I. With time and sufficient funds, NASA could develop, build and fly the Ares I successfully. The question is, should it?

First, from the perspective of schedule, the Committee observes that because of technological delays and the shortage of funds, Ares I will not effectively service the ISS, since the launch vehicle is expected to come on line in FY 2017, after retirement of the ISS. Even if the ISS is extended, within constrained resources, the Ares I and Orion will not be available until near the end of the decade, serving during only the last few years of the ISS.

As noted, safety is paramount. It is unquestionable that crews need access to low-Earth orbit at significantly lower risk than the Shuttle provides. The best architecture to assure such safe access would be the combination of a high- reliability rocket and a capsule with a launch escape system. While Ares I and Orion fit that description, so do other alternatives. The Committee was unconvinced that enough is known about the potential failures of any of the prospective high-reliability launchers plus capsule and launch escape systems to distinguish their safety in a meaningful way. The uncertainty in the safety models is large compared to the differences they predict, among competing systems, and it is clear that many of the failure modes observed in practice are not captured in the safety analysis.

The budgetary environment today is significantly more constrained than in the assumptions used for the ESAS. Despite the significant architectural commonality of the Ares I and Ares V, the program now estimates that Ares I will cost $5 billion to $6 billion to develop, even assuming that *all common costs are carried by the Ares V.* Within existing budget constraints, that will delay the availability of the Ares V to the mid-2020s if the ISS is not extended, and another several years if the ISS is extended. When it begins operations, the Ares I and Orion would be a very expensive system for crew transport to low-Earth orbit. Program estimates are that it would have a recurring cost of nearly $1 billion per flight, even with the fixed infrastructure costs being carried by Ares V. The issue is that the Orion is a very capable vehicle for exploration, but it has far more capability than needed for a taxi to low-Earth orbit.

Another understanding that has changed since the ESAS was performed is the traffic model. Figure 6.4.4-1 indicates the number of operational flights of the Ares I based on the choices made in two other decisions, and based on projected flight rates and the schedules estimated for the *less- constrained budgets.* There are no Gemini-style missions included in this count that simply orbit and do not service the ISS. In none of the combinations are there more than about six Ares I flights in the next decade, or a dozen in the next two decades.

In the years since the ESAS, other conditions have changed as well. The NASA workforce has learned from the development of Ares I. With the approaching launch of the Ares 1-X flight test vehicle, much of what will be learned may have already occurred. The sunk costs in Ares I will be partially recovered in the development of the Ares V, due to the commonality of the SRB, J2-X engine, etc. Further, a commercial space industry has continued to develop, in part due to the investment of NASA in the Commercial Orbital Transportation Services (COTS) Program. Thus, the use of commercial vehicles to transport crews to low-Earth orbit is much more of an option today than it might have been in 2005.

		Heavy-Lift Launch System Chosen	
		Ares I plus Ares V	Ares V Lite in dual mode
U.S. Participation in the ISS	End in 2015	2010-2020: 0 2020-2030: 10 or fewer	2010-2020: 0 2020-2030: 0
	Continue to 2020	2010-2020: 6 or fewer 2020-2030: 6 or fewer	2010-2020: 6 or fewer 2020-2030: 0

Source: Review of U.S. Human Spaceflight Plans Committee

Figure 6.4.4-1. Estimated total number of Ares I flights with different decisions on heavy lift and U.S. participation in ISS, assuming less-constrained budgets and no Earth-orbital flights other than those that service the ISS

Moving towards commercial crew services will also contribute to the evaluation on Economic Expansion. Together with commercial launch services for cargo to the ISS, and potentially in-space refueling, the commercial crew options could further stimulate the development of a domestic competitive launch capability. Eventually, it could stimulate a commercial service for human transport to low-Earth orbit that would be available to other markets.

In summary, the Committee found more potential contribution to the evaluation of Integrated Options due to the development of a commercial crew service to low Earth orbit than in the continued development of the Ares I. Unfortunately, neither option is without problems. The Ares I would be safe, but late to serve the ISS and expensive to operate. It would not be operated very often, or many times in total. It would delay by years NASA's start on a heavy- lift launcher. Although some of the development so far will be applicable to the Ares V, terminating Ares I would cause programmatic disruption. On the other hand, programmatic commitment at this time to commercial crew service to low- Earth orbit has benefits and risks. It has the potential to be safe, sooner and significantly less expensive. It would allow NASA to share operating costs with other customers. While the domestic development capability is demonstrated, some of the systems are largely notional. The Committee finds that if this alternative is pursued, the backup of a human- ratable heavy launcher should be accelerated, as discussed in Section 5.3.

6.4.5. Examination of the key question on shuttle extension

What should be the future of the Space Shuttle? A prudent fly-out of remaining flights (currently part of NASA policy, but FY 2011 funding for it is not part of the President's budget) or an extension of the Shuttle through 2015 at minimum flight rate? A third option, discussed in Section 4.1, of extending the Shuttle life by one flight is considered to be a variant of the 2011 fly-out option, which should be resolved by NASA.

The potential advantages of extending the Shuttle through 2015 at a low but safe flight rate are to continue to support the ISS with heavy logistics, to smooth the short-term workforce dislocation of Shuttle workforce, and to help preserve the critical workforce skills associated with launch operations. Extending the Shuttle would also help "close the gap" by delaying the retirement of the only system the U.S. currently has, or is likely to have in the next five years, to deliver humans to low-Earth orbit.

When viewed in the Integrated Options, some of these potential advantages are conditional. Extending the Shuttle in combination with developing the Ares I and Orion would not entirely close the U.S. crew-launch gap. If the Shuttle is retired in 2011, the Ares I plus Orion would become available in 2017, producing a gap of about 7 years. If the Shuttle is extended, within a fixed budget, the funds that would have paid for the development of the Ares I and Orion will be further limited, and that will delay their availability until late in the 2010s, producing a gap of at least several years at that time. Additionally, the infrastructure changes and workforce transition required for Ares I would be delayed. The gap is not closed, but shifts to the future. The only way to close the gap in U.S. crew launch is to commission a commercial service for transporting crew to low-Earth orbit—which, because it is potentially less expensive to develop, may, at some risk, be available by 2016, even with extension of the Shuttle. Other than this scenario, the Committee found no way to close the gap. The inclusion in many Integrated Options of reliance on international crew launch services is an indication that the Committee found this to be an acceptable alternative as an interim measure.

Extending the Shuttle would provide more up-mass and down-mass capability to the ISS in this interval, which would be a benefit. The current U.S. Space Transportation Policy, dated January 6, 2005, prohibits the government from taking actions that would put it in competition with commercial providers in space transportation. There is already a contract for NASA to buy commercial cargo launches, and it is not the Committee's intent that a possible Shuttle extension disrupt plans for those commercial flights. Any additional Shuttle flights would supplement the ability of the commercial carriers to service the ISS; the one integrated option that includes a Shuttle extension specifically includes the full manifest of commercial cargo flights through 2015. Extension of the Shuttle would require that the recertification done by NASA be verified, to ensure it is consistent with the CAIB recommendation. Shuttle retirement is the current NASA plan, which is a position supported by the Aerospace Safety Advisory Panel.

Extending the Shuttle would have a beneficial impact on the near-term workforce issues. Some workforce reductions would be indicated by the reduced flight rate proposed, but there would be several years in which to manage these reductions. In 2015, when the Shuttle finally retires, no NASA crew launch system would be available for several more years, and then the problem of maintaining key workforce skills would resurface. If, however, the commercial crew option were to be ready by 2016 or so, some national competence in crew launch would be nearly continuous.

Technically, extending the Shuttle makes the most sense if a directly Shuttle-derived vehicle is chosen to replace the Shuttle, which is the case in Option 4B. The relative advantages of this option are discussed below in Section 6.5.2.

Taking all factors into consideration, the decision to extend the Shuttle or not is a complex trade. Consideration of near- term access to low-Earth orbit, workforce and skills issues supports the extension. These benefits primarily materialize if the Shuttle extension is complemented by the development of commercial crew service to low-Earth orbit. The potential life-cycle costs and lower capability of the associated heavy launch system favor early retirement of the Shuttle.

6.5. Flexible Path Integrated Options Fit to the Less-Constrained Budge

6.5.1. Evaluation of integrated option 5

In the final family of options are those that pursue exploration using the Flexible Path strategy, discussed in Section 3.5. Like the Integrated Options in Section 6.4, these are not constrained to the FY 2010 budget profile. Rather, they are fit to the "less-constrained" planning budget that the Committee developed. As shown in Figure 6.2.3-1, that budget profile increases to $3 billion above the FY 2010 guidance between FY 2011 and FY 2014, and then grows at a rate comparable to an expected inflation rate of 2.4 percent per year.

Option 5. Flexible path

This option follows the Flexible Path as an exploration strategy. It operates the Shuttle into FY 2011, extends the ISS until 2020, and funds technology development. In all three variants, as shown in Figure 6.5.1-1, the commercial transport service becomes available in the mid-to-late 2010s to begin ferrying U.S. crew to the ISS. By the early 2020s, after the heavy-lift vehicle is developed, development of a small in-space habitat and an in-space restartable propulsion stage follows. All three variants also include a hybrid lunar lander that is smaller than the Altair. (See Figure 6.2.2-2.) The ascent stage is developed by NASA, but the descent stage is assumed to be commercially developed, building on the growing industrial capability pursuing NASA's Lunar Lander Challenge and the Google Lunar X-Prize. The commercial lander could also use the NASA-developed, in-space restartable engine that would be used for missions on the Flexible Path. There are three variants within this option; they differ only in the heavy-lift vehicle.

Variant 5A is the Ares V Lite variant. It develops the Ares V Lite, the most capable of the heavy-lift vehicles in this family of options. Figure 6.5.1-1 shows the schedule for this option. The Ares V Lite becomes available in the early 2020s, and the Flexible Path missions—to the Moon, Earth escape to Lagrange points or near-Earth objects, and a Mars fly-by—occur at about one-year intervals. Initial lunar landing takes place in the mid-to-late 2020s. Lunar buildup occurs at a rate of about two flights per year with the more capable Ares V Lite. On-orbit refueling, or the use of a second Ares V Lite, is necessary for the most energetic of the Flexible Path missions.

Variant 5B employs an EELV-heritage commercial superheavy-lift launcher and assumes a different (and significantly reduced) role for NASA. It has an advantage of potentially lower operational costs, but requires significant restructuring of NASA. It follows the same timeline as variant 5A up to the landing on the Moon. On-orbit refueling is used for the Flexible Path missions. Thereafter, the EELV-heritage Super Heavy flies two missions of three launches per year to the Moon, but does not carry as much load on each mission (unless on-orbit refueling is used). Thus a slower lunar development, or the development of a less massive lunar infrastructure, results.

Variant 5C uses a directly Shuttle-derived, heavy- lift vehicle, taking maximum advantage of existing infrastructure, facilities and production capabilities. It, too, follows about the same timeline up to the first lunar landing, and uses on-orbit refueling for Flexible Path missions. When lunar missions begin, the higher recurring cost of the more directly Shuttle-derived heavy launcher causes a slower rate of lunar buildup.

Comparison of the three Flexible Path valuations is shown in Figure 6.5.1-2. The distinguishing features are:

- The Ares V Lite Option 5A has an edge in Exploration Preparation, due to the more capable vehicle, and in Mission Safety Challenge because the more capable vehicle requires less complex ground and on-orbit operations.
- The EELV-heritage Super Heavy Option 5B has an edge in technology, because it includes a new U.S.- developed large hydrocarbon engine, and the lowest (i.e. best) Life-Cycle Costs, due to the commercial nature of the operation. It does poorly in Sustainability, due to the disruption in contracts, workforce transitions and the new way of doing business that would be necessary at NASA.
- The more directly Shuttle-derived heavy launcher in Option 5C has an edge in Sustainability, due to advocacy for Shuttle-derived systems, but does poorly in Life-Cycle Costs.

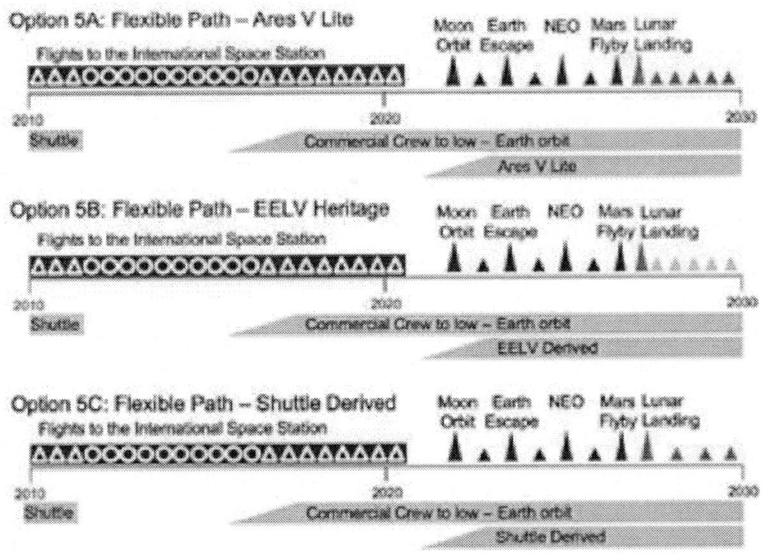

Source: Review of U.S. Human Spaceflight Plans Committee

Figure 6.5.1-1. Timelines of the three Integrated Options that follow the Flexible Path strategy, fit to the less-constrained budget. Milestones (above the line) indicate pace of activity, while ramped bars (below the line) indicate approximate range of uncertainty in the date at which the capability becomes available

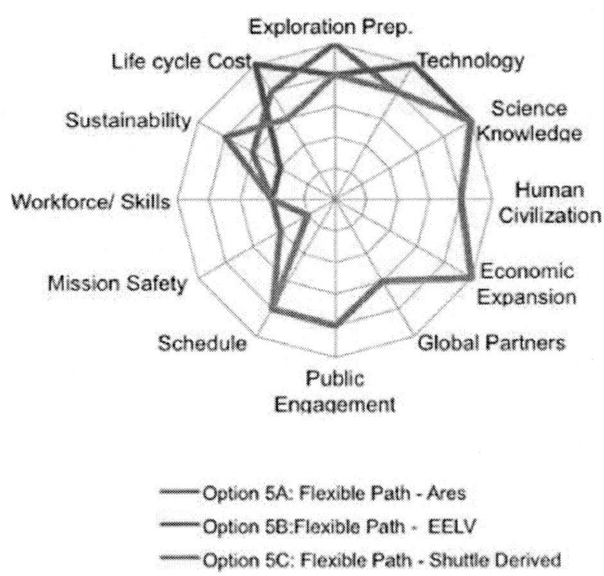

Source: Review of U.S. Human Spaceflight Plans Committee

Figure 6.5.1-2. Relative evaluation of factors in Option 5A (Flexible Path – Ares V Lite), Option 5B (Flexible Path – EELV Derived), and Option 5C (Flexible Path – Shuttle Derived), subject to the less-constrained budget. Options closer to the center are of lesser value.

6.5.2. Examination of the key question on Ares V family vs. Shuttle-derived heavy launcher

In Section 6.4.3, the decisions on heavy lift were outlined, and the comparison between the Ares V and Ares V Lite was discussed. In this section, the comparison between the Ares

V family and the more directly Shuttle-derived launcher, as indicated by the decision tree of Figure 6.4.3-1, will be examined. The background for these decisions was presented in Section 5.2.

As discussed in Section 6.4.3 above, the two Ares V family launchers under consideration are the Ares V (with 5.5-segment SRBs and six RS-68-family engines) and the Ares V Lite (with 5-segment SRBs and five RS-68-family engines). In comparison with the directly Shuttle-derived vehicles, these differences among the Ares V family are small.

As also discussed in Section 5.2, the primary candidate for consideration from the more directly Shuttle-derived family is the in-line variant. This vehicle uses two 4-segment SRBs, some number of Space Shuttle Main Engines (SSMEs) or the engine's expendable derivative, the RS-25E (three or four, typically) on the bottom of a re- engineered 8.3-meter-diameter tank, with an in-line EDS and, on top, a cargo payload or human-carrying capsule with a launch escape system. To the less-trained eye, such a vehicle would look much like an Ares V. In fact, at the end of the ESAS study in 2005, the candidate for the CaLV (Cargo Launch Vehicle) was exactly the Shuttle-derived variant just described. In the four years since ESAS, the design has evolved into the Ares V known today. Thus, the considerations for this choice of a more directly Shuttle- derived vehicle vs. Ares V more or less exactly play out the trade studies conducted by NASA in the last four years.

The principal difference between the two families is the use of the SSMEs or their expendable derivative on the more directly Shuttle-derived launcher versus the use of the RS-68-family engines on the Ares V. The use of more Shuttle-derived components lowers the development cost somewhat, and accelerates by about a year the availability of heavy lift. But these dates of first availability of heavy lift are in the early 2020s at best, due to budget contraints and likely extension of the ISS. Therefore, even if a Shuttle– derived vehicle is developed, and the Shuttle is extended, there is about a decade of gap in heavy-vehicle operations. This erodes the benefit of using the Shuttle extension and a more directly Shuttle-derived vehicle to close the critical workforce gap.

Using the more directly Shuttle-derived vehicles does produce a somewhat less-capable vehicle (Figures 5.2-2 and 5.2.1-1) and increases the recurring cost for a given mass launched to low-Earth orbit. For example, in a year of planned Constellation lunar operations in the mid-2020s, there would be three Shuttle-derived vehicle launches for each mission to the Moon, which would deliver a mass comparable to that of two Ares V-class launchers. Cargo missions would use one or two Shuttle-derived launchers. With two crew and two cargo missions per year, this would require eight to ten launches of the Shuttle-derived launcher, each with three or four SSMEs or derivatives, for a total of 24 to 40 of the Shuttle engines being used, with a resulting high recurring cost.

Among the other notable differences between the Ares V family and the more-directly Shuttle-derived launcher family is the mission-launch reliability. Since the latter requires three launches for each planned Constellation lunar mission, there would be a somewhat lower reliability in any given time window than would be provided by the Ares V, which only would require two launches in the same time window.

The Committee considered as an issue the commonality with the national space industrial base. The Ares V uses engines from the RS-68 family, with commonality in the industrial base with those used on the EELVs by National Security Space. Both the Ares V and the more-directly Shuttle- derived vehicle have commonality in the solid-rocket motors with vehicles used in National Security Space.

In summary, the Committee viewed the decision between the Ares V family and the Shuttle-derived family as one driven by cost and capability. The development cost of the more Shuttle-derived system would be lower, but it would be less capable than the Ares V family and have higher recurring costs. There are potential workforce and skill advantages associated with the use of the more-directly Shuttle-derived system, but the long gap between when the Shuttle is retired in 2011, or even 2015, and when the Shuttle-derived heavy-lift launcher becomes available in the early to mid- 2020s would diminish the potential value of the workforce continuity associated with Shuttle derivatives.

6.5.3. Examination of the key question on NASA heritage vs. EELV-heritage super-heavy vehicles

The highest-level decision on heavy-lift-launch vehicles is whether to base the launch system for exploration on these NASA-heritage vehicles or on the further extension of the EELV-heritage vehicles up to the 75- mt range (Figure 6.4.2-1). It should be emphasized that this is not the existing EELV heavy launcher, which has a maximum payload to low-Earth orbit of about 25 mt, but rather requires the development of a substantially new vehicle, in part based on existing components and manufacturing facilities.

The EELV-heritage super-heavy launch vehicle would be capable of launching about 75 mt to low-Earth orbit, significantly less than the Ares V family at 140 to 160 mt, or the Shuttle-derived vehicle in the range of 100 to 110 mt. However, the EELV-heritage super heavy is still larger than the Committee's estimated smallest possible launcher to support exploration, which is in the range of 40 to 60 mt. In a nominal piloted lunar mission, without in-space refueling, the EELV-heritage super heavy would require three flights, potentially with an additional crew taxi flight, versus two launches for the Ares V Lite. In the Flexible Path options, the EELV would not require more launches, but would involve more on-orbit operations than the Ares V-family approach. For these launch rates, the EELV would have a lower recurring cost than the NASA-heritage vehicles.

Initially, the EELV-heritage super heavy vehicle would use the Russian RD-180 hydrocarbon fueled engine, currently used on the Atlas 5. In the cost analysis utilized by the Committee, provision was made for the development of a new large domestic engine to replace the RD-180 for both NASA and National Security missions. This would have technology benefits, and would provide value to National Security systems.

While there are technical differences between the two families, the Committee intended the principal difference to be programmatic. The EELV-heritage super heavy would represent a new way of doing business for NASA, which would have the benefit of potentially lowering development and operational costs. The Committee used the EELVheritage super-heavy vehicle to investigate the possibility of an essentially commercial acquisition of the required heavy- launch capability by a small NASA organization similar to a system program office in the Department of Defense. It would eliminate somewhat the historic carrying cost of many Apollo- and Shuttle-era facilities and systems. This creates the possibility of substantially reduced operating costs, which may ultimately allow NASA to escape its conundrum of not having sufficient resources to both operate existing systems and build a new one.

However, this efficiency of operations would require significant near-term realignment of NASA. Substantial reductions in workforce, facilities closures, and mothballing would be required. When the Committee asked NASA to assess the cost of this process, the estimates ranged from $3 billion to $11 billion over five years. Because of these realignment costs, the EELV-heritage super heavy does not become available significantly sooner than the Ares V or Shuttle-derived families of launchers. The transition to this way of doing business would come at the cost of cutting deeply into a the internal NASA capability to develop and operate launchers, both in terms of skills and facilities.

There would be other consequences at the national level. Needless to say, the co-development of the EELV-heritage super heavy would require careful coordination between NASA and the Department of Defense to ensure joint value.

In summary, the Committee considers the EELV-heritage super-heavy vehicle to be a way to significantly reduce the operating cost of the heavy lifter to NASA in the long run. It would be a less-capable vehicle, but probably sufficiently capable for the mission. Reaping the long-term cost benefits would require substantial disruption in NASA, and force the agency to adopt a new way of doing business. The choice between NASA and EELV heritage is driven by potentially lower development and operations cost (favoring the EELV-heritage systems) vs. continuity of NASA's system design, development and mission assurance knowledge and experience, which would provide higher probability of successful and predictable developments (favoring NASA systems). EELV-heritage launch systems, due to their lower payload performance, would require significantly greater launch and mission complexity to achieve the same total mass in orbit. The EELV option would also entail substantial reductions in the NASA workforce and closure of facilities necessary to obtain the expected cost reductions.

6.6. Comparisons across Integrated Options

6.6.1. Cross-option comparisons

A cross-family comparative evaluation is shown in Figure 6.6.1-1, which contrasts the Baseline (Option 3) with the Ares V Lite variant of the Flexible Path (Option 5A). The Flexible Path option scores more highly than the Baseline on 9 of the 12 criteria. The higher rankings include:

- Exploration Preparation (due to much more capable launch system)
- Technology (due to investment in technology)
- Science (because of more places visited)
- Human Civilization (due to the ISS extension)
- Economic Expansion (because of commercial involvement in space elements and crew transport)
- Global Partnerships (gained by extending the ISS)
- Public Engagement (by visiting more new locations, and doing so each year)
- Schedule (exploring beyond low-Earth orbit sooner)
- Life-Cycle Costs (due to commercial crew services)

Source: Review of U.S. Human Spaceflight Plans Committee

Figure 6.6.1-1. Relative evaluation of factors in Option 3 (Baseline - Program of Record) and Option 5A (Flexible Path - Ares V Lite), subject to less- constrained budget. Options closer to the center are of lesser value

6.6.2. Examination of the key question on exploration strategy

The fifth and final of the key questions guiding the decisions on the future of human spaceflight is: What is the most practicable strategy for exploration beyond low-Earth orbit? In Chapter 3, three exploration strategies were presented, but the choice of exploration of Mars First was found not to be viable at this time. The two remaining choices are:

- Moon First on the Way to Mars, with surface exploration focused on developing capability for Mars
- Flexible Path to Mars via the inner solar system objects and locations, with no immediate plan for surface exploration, then followed by exploration of the lunar and/or Martian surface

This contrast can be highlighted by comparison of the reference variant for Moon First (Option 4A) and Flexible Path (Option 5A), as shown in Figure 6.6.2-1. These two options differ only in exploration strategy. The Flexible Path equals or exceeds the ratings of the Moon First option in all areas. It has an advantage in: Science, Economic Expansion, Public Engagement and Schedule. These four distinctions will be examined below.

From the perspective of Science Knowledge, the Moon First approach would allow better understanding of the evolution of the Moon, and use the Moon's surface as a record of events in the evolution of the solar system. The Flexible Path would explore near-Earth objects, and also demonstrate the ability to service science observatories at the Lagrange points. The crews on such missions would potentially interact with robotic probes on the surface of Mars, returning samples as well. In some of its alternatives, including the one costed in Section 6.5, the Flexible Path also allows exploration of the Moon, though at a more limited scale than in the Moon First approach. Considering that we have visited and obtained samples from the

Moon, but not near-Earth objects or Mars, and also that the Flexible Path develops the ability to service space observatories, the Science Knowledge criterion slightly favors the Flexible Path. Broadly, the more complex the environment, the more astronaut explorers are favored over robotic exploration. In practice, this means that astronauts will offer their greatest value-added in the exploration of the surface of Mars.

It is likely that the Flexible Path approach would engender more Public Engagement than the Moon First approach. In every flight, the Flexible Path voyages would visit places where humans have never been before, with each mission extending farther than the previous one, potentially leading to a full dress rehearsal for a Mars landing. A potential liability of the Moon First option is that it could appear to some stakeholders as a modern repetition of exploration that was accomplished 50 years earlier.

Schedule also favors the Flexible Path scenario. The fundamental economics of the investment by NASA to begin flights on the Flexible Path and Moon First options are shown in Figure 6.6.2-2. Before lunar exploration can begin, NASA must complete four development programs: the heavy-lift launcher, the Orion capsule, the Altair lander, and at least some of the lunar surface systems. Even the well-funded Apollo Program only had to complete the first three of these. In contrast, exploration on the Flexible Path could begin with just the capsule and launcher, and then slowly develop much less costly in-space propulsion stages and habitats. After NASA explores on the Flexible Path for a half decade or so, it could then invest in the lunar lander and surface systems. In summary, the Flexible Path provides for exploration beyond low-Earth obit several years earlier, and allows a less demanding programmatic investment profile.

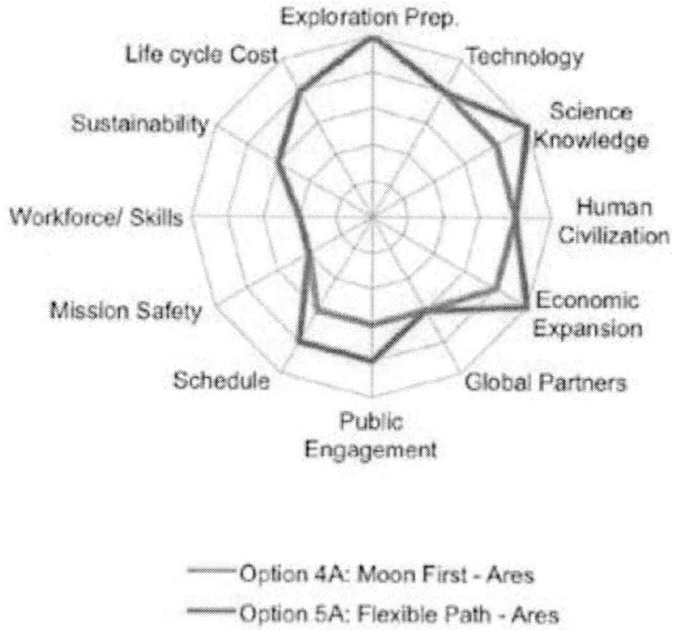

Source: Review of U.S. Human Spaceflight Plans Committee

Figure 6.6.2-1. Relative evaluation of factors in Option 4A (Moon First – Ares V Lite) and Option 5A (Flexible Path – Ares V Lite), subject to the less- constrained budget. Options closer to the center are of lesser value

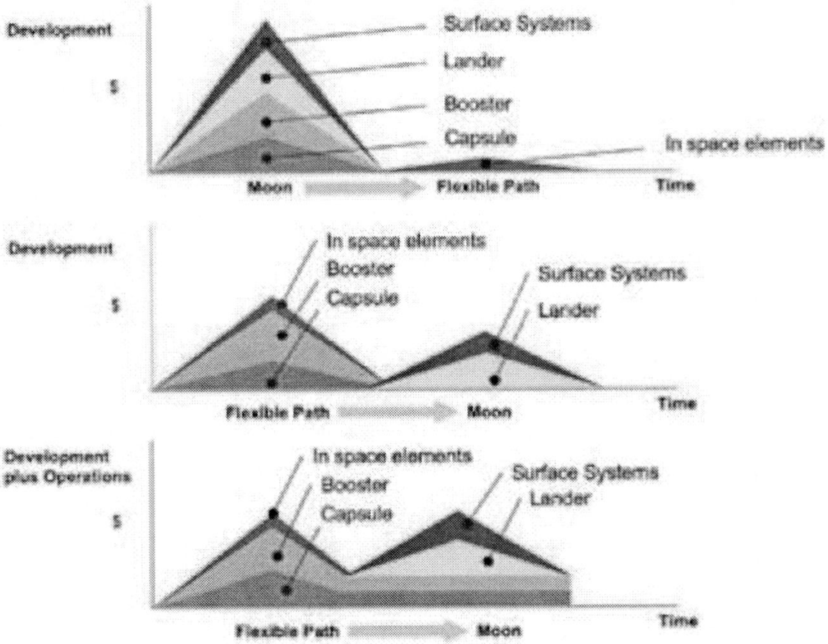

.Source: Review of U.S. Human Spaceflight Plans Committee

Figure 6.6.2-2. Notional development cost (top and middle) and total program costs (development and operations - bottom) suggesting that the phasing of Flexible Path followed by the Moon First produces a more level budgetary profile

Because the Flexible Path option contained a commercially developed lunar lander descent stage, it was evaluated more highly in Economic Expansion as well. The use of a commercial lander is not fundamental to the execution of the Flexible Path, but is more likely in this strategy. The lunar landing would be later, involve a simpler lander, and follow the development by NASA of the in-space re-startable engine, all of which would make a commercial system more viable in the Flexible Path than in the Moon First strategy.

Of the evaluation criteria on which the two strategies score equally, there are some distinctions. Under Human Civilization, both lead to better understanding of human adaptation to space, but the Flexible Path aids in the protection of Earth from near-Earth objects. From the viewpoint of Mission Safety Challenges, the two strategies are also about equal. Operations at the Moon are closer and allow return to the Earth more rapidly, but landing on and launching from a surface is a dynamic environment. In contrast, the Flexible Path missions are less dynamic, but occur farther from Earth.

There is no reason to believe that the remaining evaluation criteria favor one or the other strategy for exploration. They have more to do with how the strategy is implemented. For example, either the Moon First or Flexible Path could be the basis for a new or extended international partnerships in space.

In summary, 8 of the 12 criteria favor neither the Moon First nor Flexible Path strategies for exploration. However, of the four that do discriminate—Science Knowledge, Public Engagement, Schedule and Life- Cycle Costs—all slightly to moderately favor the Flexible Path.

Finding on Evaluation of Program Options

Options for human spaceflight should be evaluated by a set of criteria that are consistent with goals. The Committee identified 12 criteria which measure the capability of an option to satisfy its stakeholders, along the motivations listed above, along with programmatic issues of safety, cost, schedule, sustainability and workforce impact. It is the role of decision-makers to prioritize these measures.

Findings on Options for the Human Spaceflight Program

The Committee developed five alternatives for the Human Spaceflight Program. In reviewing these, it found:

- Human exploration beyond low-Earth orbit is not viable under the FY 2010 budget guideline.
- Meaningful human exploration is possible under a less-constrained budget, ramping up to approximately $3 billion per year in real purchasing power above the FY 2010 guidance in total resources.
- Funding at the increased level would allow either an exploration program to explore the Moon First or one that follows a Flexible Path of exploration. Either could produce results in a reasonable timeframe

7. Critical Technologies for Sustainable Exploration

The Space Act of 1958 calls on NASA to preserve the U.S. position as a leader in space technology. Today, the alternatives available for exploration systems are severely limited because of the lack of a strategic investment in technology development in past decades. Looking forward, NASA has before it an unprecedented opportunity to make an effective strategic technology plan. With a 20-year roadmap of exploration laid before it, NASA can make wise technology investments that will enable new approaches to exploration. Two recent reports of the National Academies have made recommendations in this regard.

The investments should be designed to increase the capabilities and reduce the costs of future exploration. NASA has conducted studies to demonstrate the mass reduction, and therefore operational cost savings, that are achievable with investments in technology. (See Fig 7.1-1.) As an illustration, it indicates an almost ten-fold reduction in mass required in mass required for future missions to Mars. If appropriately funded, a technology development program would re-engage the minds at American universities, in industry and within NASA. This will benefit human and robotic exploration, the commercial space community, and other U.S. government users alike.

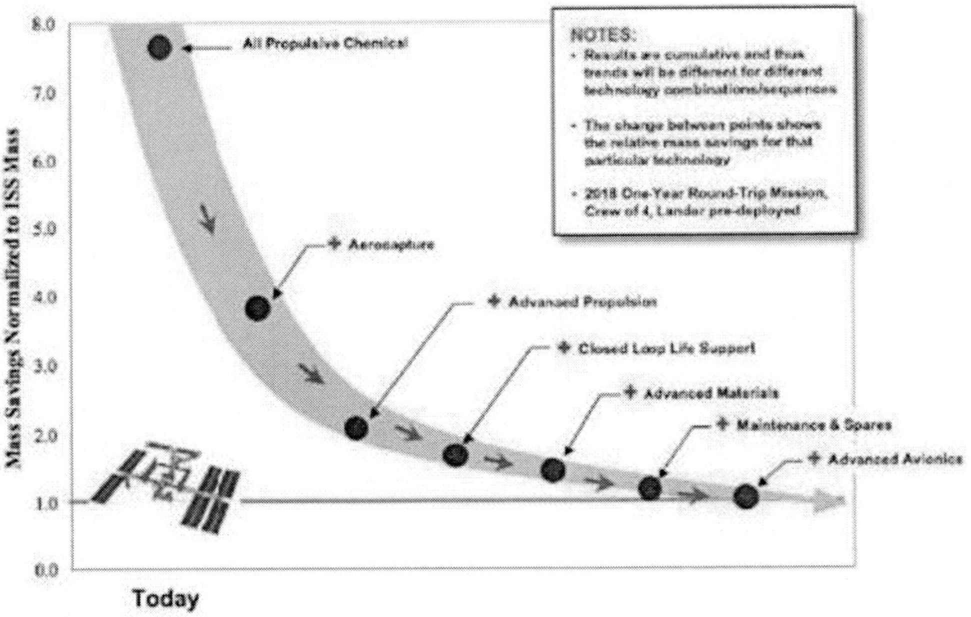

Source: NASA

Figure 7.1-1. With technology investments, the mass required for a Mars exploration mission decreases from eight times the mass of the International Space Station to a mass comparable to the Station

7.1. Fundamental Unknowns

Three factors affecting long-duration human space exploration are of central importance, yet do not lend themselves to definitive assessment based on the available data: (1) the effects of prolonged exposure to solar and galactic cosmic rays on the human body; (2) the impact on humans of prolonged periods of weightlessness followed by a sudden need to function, without assistance, in a relatively strong gravitational field; and (3) the psychological effects on individuals facing demanding tasks in extreme isolation for well over year with no possibility for direct outside human intervention.

While the specific technologies to address needed capabilities can and will be debated, if the United States wishes to conduct more and more capable missions in the future with nearly constant budgets, it is essential to develop and bring to flight readiness the technologies required. This will not happen without a sustained plan in which needed capabilities are identified, multiple competing technologies to provide that capability funded, and the most mature of them demonstrated in flight so that exploration architectures can then depend on them. For many technologies, it is less expensive to design a flight demonstration using the facilities on ISS than it would be to design a free-flying mission for each and every technology demonstration.

Radiation effects on humans

Beyond the shielding influence of the Earth's magnetic field and atmosphere, ionized atoms that have been accelerated to extremely high speeds in interstellar space fill the solar system. The effects of such galactic cosmic radiation on crews on long-duration spaceflight far from the Earth are a significant concern. Additionally, normal solar flare activity also occasionally releases radiation potentially injurious to humans. On the Moon or on the surface of Mars, techniques are available to shield a human habitat from these sources of radiation, but the massive shielding is cost-prohibitive for a spacecraft. These radiation effects are insufficiently understood and remain a major physiological and engineering uncertainty in any human exploration program beyond low-Earth orbit. A 2008 report by the National Research Council concluded, "Lack of knowledge about the biological effects of and responses to space radiation is the single most important factor limiting the prediction of radiation risk associated with human space exploration." A robust research program in radiobiology is essential for human exploration. Research on these radiation effects on humans is limited on the ISS, since it is partly shielded by the Earth's magnetic field.

In addition to studying the effects of galactic cosmic radiation (GCR), there are mitigation strategies, such as more effective shielding techniques, or the use of high shielding mass on a reusable habitat that cycles between Earth and Mars without being accelerated for each mission (cycler habitat), that need additional study.

Micro- and hypo-gravity effects on humans

While significant data on crew adaptation to micro-gravity now exists from extended ISS stays, there is a need to further develop countermeasures and build an understanding of the even longer profiles that will be encountered in exploration beyond low-Earth orbit. The ISS is a logical place to conduct such research, and the U.S. should obtain these data before the ISS is retired.

Psychological effects of extreme isolation on humans

While many experiments have been conducted on Earth to examine the effects of prolonged isolation on humans needing to continue to function at a high level, these experiments generally fall short of simulating the circumstance of extreme physical confinement in which the participant realizes there is no opportunity to "end the experiment." In this regard, missions to Mars would be far more demanding than those to the Moon. Mars, at its closest, is 56 million kilometers from the Earth, whereas the Moon is 380,000 kilometers. In the latter case, return is generally possible within a few days. In contrast, Martian circumstances may require many months or years for an emergency return to Earth.

7.2. Propellant Storage and Transfer in Space

Wernher von Braun wrote of the significant benefits to be gained from propellant transfer and storage in space. Up to this time, the normal approach for inserting payloads on trajectories away from low-Earth orbit towards the Moon or Mars is to use an upper stage called an Earth Departure Stage (EDS). In the conventional scheme, the Earth departure stage burns some of its fuel on the way to orbit and arrives at low-Earth orbit partially full. The remainder of the fuel is expended injecting the payload toward its destination beyond low-Earth orbit. An alternative, discussed in Chapter 5, is to re-fuel in space, so that the EDS can arrive in orbit mostly empty, and be refilled. After leaving Earth, exploration systems will still need to make one, two or three propulsive maneuvers, often after months in space. Again past experience is to use storable propellants for these maneuvers, but in human exploration, the cost of doing this becomes large.

The benefit of in-space cryogenic transfer and storage is that it enables refueling in space and the use of high-energy fuels for in-space propulsion. Using propellant transfer for the EDS, for example, allows more mass to be injected from the Earth with a given launch vehicle, or a smaller launcher for a given payload.

Today, these technologies are considered ready for flight demonstration, according to both NASA and industry experts working in the field. (See Fig 7.2-1.) Nonetheless, legitimate questions remain as to the practical feasibility of the approach. These concerns generally center on cryogenic transfer and storage technology. Cryogenic transfer, storage and gauging in a micro-gravity environment create challenges that have been investigated by researchers in the laboratory for decades. Automated rendezvous and docking of delivery tankers has recently been demonstrated. Capabilities that remain to be demonstrated include:

- Long-term storage of very cold (cryogenic) propellant without excessive boil-off
- Transferring cryogenic propellant between tanks in a zero-g environment
- Making cryogenic fluid line connections
- Gauging the quantity of propellant in the tanks in a zero G environment

Figure 7.2-2 shows the maturity of various technologies based on the NASA Technology Readiness Level (TRL) system, summarizing the current assessment of capabilities required for propellant transfer and storage on orbit.

These cryogenic technologies have reached the point where a flight demonstration is the next logical step in development before this capability can begin to be designed into systems. The two technologies of liquid acquisition, collecting fluid in micro-gravity near the point from which it will be withdrawn from a tank, and automated fluid coupling cannot be tested in a realistic environment without in-space demonstration. Cryogenic coolers are needed for true zero-boil indefinite storage of liquid hydrogen (LH2), but other technologies listed in the table are intended to negate this problem for storage up to one year.

Technology	Most Mature Approach	Status
Liquid Acquisition	Surface tension device +settling	Ground Demonstrations; Needs Flight Test
Propellant Gauging	Capacitance probe +settling	Mature Technology with Settling
Ullage Gas Management	Settled vent	Mature Technology with Settling
Transfer Line Conditioning	Recirculation	Mature Technology with Settling
Automated Fluid Coupling	Extension of probe used for propellant couplings on Orbital Express	Flight Demonstrated with Storables; Needs Flight Test with Cryogens
Insulation & Thermal Management	MLI is mature technology, vapor-cooled shield further lowers boil-off	Ground Demonstrations; Needs Flight Test

Source: Review of U.S. Human Spaceflight Plans Committee

Figure 7.2-2. Technology Readiness of Components for In-Space Refueling

7.3. *In Situ* Propellant Production and Transport

After mastering the technologies for storing and transferring propellant in space, the next step is to manufacture propellant from resources already there. The lunar surface and some near-Earth objects are the only known sources of suitable resources that can be brought feasibly to cislunar space, such as to the Earth-Moon Lagrange Point L1, which was identified as a promising candidate for a cislunar propellant depot in a 2004 NASA study.

In situ propellant production requires the combination of two unique capabilities: (1) producing the propellant, and (2) transporting it economically. Oxygen is abundant in all lunar rocks and regolith (dirt without organic material), and a variety of chemical processes to extract it have been demonstrated in Earth-based laboratories. These propellant production methods must next be demonstrated on the lunar surface though robotic missions. Collecting lunar material and bringing it to a lunar-based processing station presents a great challenge. Laboratory work has shown that this will likely require both robotic and human-tended missions to mature the technology.

In addition, it is also extremely important to produce hydrogen for fuel in space. If hydrogen can be economically extracted from the Moon, it will likely serve as a source of propellant for future exploration missions. If it cannot be produced, the case for exporting propellant from the Moon becomes less compelling, and near- Earth objects would rise in importance. Discoveries in recent years do suggest the availability of significant hydrogen deposits at the lunar poles. In addition, Apollo samples showed useful hydrogen deposits from the solar wind implanted in the regolith. These regolith deposits would require processing large amounts of material, comparable to coal extraction on Earth. Robotic exploration of the Moon continues today, and further interesting sources of resources may well be found.

The composition of near-Earth objects is less well understood than the composition of the Moon because of the current dependence on telescope-based observations and inferences drawn from meteorites reaching Earth. These data suggest that almost any desirable resource can be found on near-Earth objects, but since each near-Earth object is a distinct body with its own orbit and properties, it is difficult to make generalizations about how resources would be extracted and returned to cislunar space. It is worth noting that in some cases, the energy required to return mass from a near-Earth object to near-Earth space is significantly less than to return mass from the lunar surface to Earth Moon L1 Lagrange Point. Therefore, further robotic exploration and human-tended pilot visits to near-Earth objects are particularly interesting subjects for future exploration.

Reusable chemical rockets might also be used to deliver *in situ* propellant from the Moon to the Earth Moon L1 Lagrange point. Space tugs carrying and using hydrogen are generally more compelling than tugs carrying and using other propellants. As an alternative, because of the Moon's low gravity, non-chemical propulsion should also be considered. Catapults, tethers and beamed energy are likely to become practicable for Lunar-to-L1 transport long before they become practicable for Earth-to-orbit launch, and should continue to be investigated. Nonetheless, their application appears to be far off.

Because about two-thirds of the mass on an Earth-to-Marsto-Earth mission would be propellant, cost-effective lunar- produced propellant could decrease the mass that must be lifted from Earth by a factor of two to three. Further, achieving industrial levels of oxygen and hydrogen production on Mars would greatly simplify the challenge of transporting fuel for the return trip from Mars to Earth.

In situ propellant produced on Mars has been consid- ered as well. Oxygen could be extracted from the carbon-dioxide-based Martian atmosphere, and both oxygen and hydrogen could be extracted from Mars' ample ice. Although laboratory work for extracting resources from Mars is promising, the technology remains to be demonstrated under realistic circumstances.

7.4. Mars Orbit to Surface Transportation

The entry, descent and landing of cargo on Mars is difficult because Mars has sufficient atmosphere to drive the design of landing systems, but inadequate atmosphere for feasible parachutes or wings to safely land astronauts on the surface. Scientific probes landing on Mars have used a complex mix of aerodynamic braking and rocket propulsion. These techniques will have to be improved before larger robotic or crewed missions can be sent to Mars. This research and technology development program needs to be started soon, because it will require many iterations and increasingly larger missions before NASA is ready to demonstrate a safe, crewed Mars landing. Meanwhile, the intermediate results would greatly benefit future robotic missions.

Because of the unique landing challenges posed by Mars, a robust human presence will require an advanced Mars orbit to Mars transportation system, most likely a reusable system that could transport cargo and people between the Martian surface and a depot located nearby. Nuclear thermal rockets, using *in situ* Martian-produced propellant, would fly to Martian orbit, collect a payload, and then use aerodynamic braking for the initial descent, followed by nuclear rocketry to land. Alternately, a chemical rocket would need to refuel on both Mars and in near-Mars space, such as on the Martian moon Phobos. Use of Phobos for propellant production would benefit transportation both to and from the Martian surface and provide the propellant for astronauts to return from Mars to Earth.

A Phobos-based teleoperated exploration of the Martian surface, returning with samples from that surface, would likely precede a crewed Mars landing mission, and would provide dramatically more responsive remote control than with the communication delays incurred between Mars and Earth. The use of Phobos- and/or Mars-produced *in situ* propellant could likely reduce the flight cost of a crewed Mars landing expedition by a factor of two to three.

7.5. Advanced Space Propulsion

Since the 1950s, advanced space propulsion has been recognized as an extremely desirable technology for Mars missions, as is the possibility of aerocapture. (See Fig 7.1-1.) The application of advanced space propulsion to crewed missions could significantly reduce the amount of propellant used, while providing sufficient thrust to build up interplanetary velocities relatively quickly—over days or weeks instead of months or years. Unfortunately, the physics of the problem are such that the more a vehicle tries to minimize propellant usage, the more power it consumes, yet since the power-supply weight increases with power generation requirements, a more substantial power supply can negate some or all of the benefits of lower propellant usage.

Two promising advanced space propulsion technologies are based on solar and nuclear energy sources.

- **Solar:** Current solar-power collectors are too heavy to deliver dramatic benefits in space propulsion. Solar- electric thrust is used today for some satellites and it may play a role in cargo transportation from LEO to, for example, L1 and back. Unfortunately, the acceleration it provides is so low that it would take months or years to get from Earth to the Moon, for example. As a result, for solar-powered advanced propulsion to provide revolutionary benefits, either far lighter, thin-film solar arrays must be matured, or the heavy solar power collector must be left off the space ship and the power from the collector beamed by lasers to lightweight collectors aboard the ship.
  Theoretical studies indicate this latter technology is physically plausible, but it hasn't made it out of the laboratory. In addition, the beam can only be held together over distances up to a few hundred thousand kilometers with a reasonable-sized transmitter. Solar technology requires research and development and flight demonstrations, as well as a large investment to build the solar-power collector and beam transmitter in space. However, such an effort might pave the way for transmission of power to the Earth for domestic use, although it is not known whether such systems will prove economical even if they were to be technically feasible.

- **Nuclear:** A more mature technology, nuclear-thermal propulsion reached advanced ground-firing demonstrations during the 1960s before the program was cancelled. With nuclear-thermal propulsion a nuclear reactor heats hydrogen and then ejects it to provide thrust. A nuclear-electric thruster, on the other hand, produces electricity to run an electric thruster, such as high power Hall effect thrusers or the variable specific impulse magnetoplasma rocket (VASIMR) thruster NASA is currently funding.
- Nuclear reactors today provide a substantial fraction of U.S. electricity production and other countries use nuclear reactors to produce an even larger fraction of their domestic power. For advanced space transportation, much lighter nuclear reactors are required. Although expensive to develop, this solution could cut the cost of missions from Earth by a factor of two or three. Alternately, it could be used to increase vehicle velocity for the same cost. Nuclear propulsion is probably essential for any crewed activity beyond Mars.

The space nuclear program is an excellent candidate for a multinational research effort because different countries have different capabilities and research interests. NASA would benefit from a coordinated multi-national research effort in this area.

7.6. Technology Summary

In order to give future designers a rich and effective set of technologies to draw from, an investment in a broad-based space technology program is prudent. This should be done in a focused but long-term manner, with a clear metric of enabling and reducing the cost of future exploration. There are a number of potential technologies and approaches to be examined, as indicated in Figures 7.6-1 and 7.6-2, which attempt to identify near- and longer-term benefits from the investment. Some of these technologies have been discussed above, and others throughout the report. NASA would not be the only beneficiary of these technologies. Other U.S. government and commercial users of space would benefit as well in terms of new capabilities or reduced cost. Consistent with administration planning, the Earth-based benefits to economic recovery, energy technology, biomedical science and health, and protection of our forces and homeland have been indicated.

Finding on Technology Development

Technology development for ex-ploration and commercial space

Investment in a well-designed and adequately funded space technology program is critical to enable progress in exploration. Exploration strategies can proceed more readily and economically if the requisite technology has been developed in advance. This investment will also benefit robotic exploration, the U.S. commercial space industry, the academic community and other U.S.- government users.

Capability	Approaches	Benefit to NASA	Other Benefits
Improved LEO launch	Higher volume production of expendables, or reusables	Lower cost to LEO	NSS, ComS Econ, Energy, Protect
Galactic Cosmic and Solar Radiation mitigation	Better risk assessment, mitigation shielding	Lower mass systems with better human protection	Bio
Autonomous Rendezvous and Docking	Improved sensors, algorithms, system tests	Enables un-piloted assembly and fueling in orbit	NSS, ComS Econ, Protect
Cryogenic fuel transfer	Development and testing of fuel management and transfer technology	Enable smaller earth departure stages or makes planned ones more capable	NSS, ComS Econ, Protect
In-space re-startable engine and boil-off technology	Cycle and engine development of hydrogen or methane engines	Enables longer duration exploration to destinations needing return propulsion	NSS, ComS Econ, Energy, Protect
Crew autonomy and robotic support	Development of robotics to assist crew in space and on surfaces	Increases efficiency and focus of crew on tasks better done by humans	Bio, Protect
Advanced system design methodologies	Computational, simulation and decision support systems	Quicker, better documented and more thorough system designs	NSS, ComS Econ, Energy, Protect
Advanced Materials	High temperature and nano engineered	Improves durability and strength, lighter weight systems	NSS, ComS Econ, Energy, Bio, Protect
Closed-loop life support	Mechanical or biological systems	Lower mass for human duration	ComS Econ, Bio, Protect
Methane engine	Cycle and engine development of methane engines	Allows better packaged stages, that would use Mars ISRU products	NSS, ComS Econ, Energy, Protect
Aero-entry, descent and landing in atmospheres	Development of novel structures for aero-braking and entry	Lowers mass of systems, enables large masses on Mars	NSS, ComS Econ, Protect

Technologies are listed in an approximate order of decreasing importance and urgency
Other Benefits are to National Security Space, Commercial Space, Economic recovery, Energy technology, Biomedical science/health and Protection

Source: Review of U.S. Human Spaceflight Plans Committee

Figure 7.6-1. Technology opportunities to impact near- and mid-term exploration capabilities and sustainability

Capability	Approaches	Benefit	Other Benefits
Surface systems	Assembable or inflatable habitats and mobility systems	Lower mass to surface, more capable exploration	ComS Econ, Protect
Surface power	Advanced solar, isotope or fissile based	Lower mass to surface, more power for exploration and ISRU	NSS Econ, Energy, Protect
Lunar In Situ Resource Utilization	Regolith or "hydrogen" rich formations derived	Lower consumable delivery to surface	ComS Econ
Lunar In Situ Propellant Production & export	Reusable lunar to EM L1 tug	Enabling of new Earth-Moon-Mars transportation architectures	ComS Econ
NEO based In Situ Resource Utilization	Based on initial sample assays and transportation network	Enabling of new inner solar system transportation architectures	ComS Econ
Advanced in-space propulsion	Beamed-power or nuclear	Smaller in-space propulsion systems, significantly lower mass to LEO	NSS, ComS Econ, Energy, Protect
Mars In Situ Propellant Production	Atmosphere or ice derived	Lower mass for human duration and enabling of lighter ascent vehicles	ComS Econ
Earth Mars cycling spacecraft	Aldrin and Multi-synodic cyclers	Enabling of new mode of interplanetary travel	

Technologies are listed in an approximate order of importance and urgency
Other Benefits are to National Security Space, Commercial Space, Economic recovery, Energy technology, Biomedical science and Protection

Source: Review of U.S. Human Spaceflight Plans Committee

Figure 7.6-2. Technology opportunities impact longer-term capabilities and sustainability

8. PARTNERSHIP

Various forms of partnerships have been discussed throughout the report. This section examines in greater detail international partnerships (Section 8.1), as well as partnerships among elements of the U.S. government (Section 8.2).

8.1. International Partnerships

The human exploration of space is historically intertwined with the recent evolution of America's international relationships. The U.S. has moved from an era of competing with the Soviet Union in the Apollo era, to collaborating with our historical allies in Space Station *Freedom*, to embracing the Russians in the International Space Station, and now to engaging many potential new partners.

Space exploration has become a global enterprise. Many nations have aspirations in space, and the combined annual budgets of the space programs of our principal partners are comparable to NASA's. (See Figure 8.1-1.) If the United States is willing to lead a global program of exploration, sharing both the burden and benefit of space exploration in a meaningful way, significant accomplishments could follow. Actively engaging international partners in a manner adapted to today's multi-polar world could strengthen geopolitical relationships, leverage global financial and technical resources, and enhance the exploration enterprise.

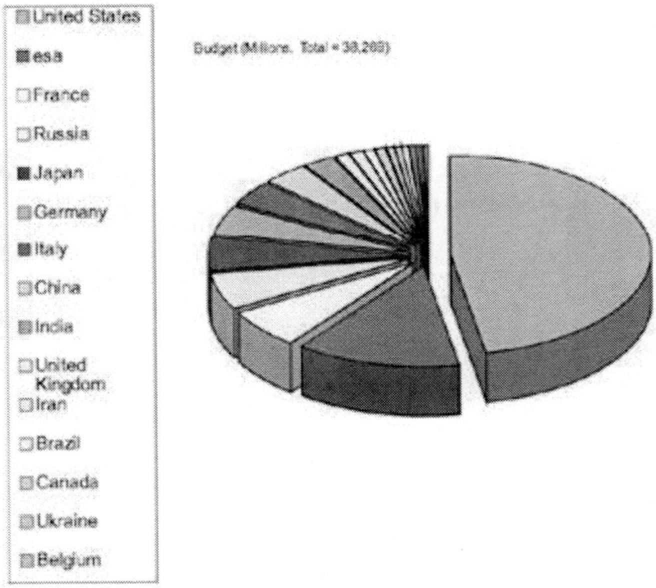

Source: Various Sources

Figure 8.1-1. The annual budgets of civil space agencies in 2007. The sum of the budgets of potential partners is now comparable to NASA's budget

One means of reducing the funding demands of major human spaceflight programs is to join in partnerships with other nations that share common space goals. Thus far, three nations have by themselves placed astronauts in space: the U.S., Russia and China. International programs offer the added advantage of providing access to advanced technology not available in the U.S., an increasingly common circumstance (e.g., Russian-designed, hydrocarbon-(RP-1)- fueled liquid rocket motors). Such arrangements also facilitate cost sharing.

The principal disadvantage of international programs (ex- cluding business-to-business arrangements based on en-forceable contracts) is that nations are sovereign entities and, as such, can unilaterally change their plans—which can be very disruptive. Much of the international community, probably justifiably, faults the U.S. with regard to this practice. But perhaps an even greater impediment to U.S. involvement in international cooperative programs is the U.S. International Trafficking in Arms Regulations (ITAR). The Committee deems these laws to be outdated and overly restrictive for the realities of the current technological and international political environment.

International programs are generally more difficult to manage than national programs, in part because of the need for a greater degree of consensus and coordination; however, this can often be overcome when one of the participating nations serves in a leadership role—as distinct from a domineering role. The management structure that has evolved for the ISS has proven particularly effective and could serve as the basis for the next major international cooperative human venture into space. The Committee suggests that the President invite America's international partners to join with America in discussions of this possibility.

The question occasionally arises whether the public would support space accomplishments shared among nations, as opposed to those representing purely "American" achievements. But the fact is that human spaceflight has always existed in an international context, first as competition and more recently as cooperation. It is difficult to argue against the latter. Moreover, the U.S. space program, at its height, had the self-confidence to emphasize that the United States was acting as a representative of all humanity. The Apollo 11 plaque left on the Moon reads: "Here men from the planet Earth first set foot upon the Moon July 1969, A.D. We came in peace for all mankind."

If international programs are to succeed (i.e., where a true partnership that benefits all parties exists), there will inevitably be some measure of dependency of the member nations upon one another. In some cases this may imply dependency with regard to capabilities that lie on a mission-critical path—witness the impending launch gap. Nonetheless, if effective and meaningful partnerships are to be established, some degree of such dependency is simply the price of admission—a price, the Committee concludes, that is worth paying in most instances.

International cooperation has been fruitful for both robotic and human space missions. The U.S. has enjoyed long, positive space exploration partnerships with several countries. The U.S. has participated in cooperative efforts with Europe for many years, both through independent European country agencies and through the European Space Agency. Similarly, the U.S. has excellent historical space relationships with Japan and Canada.

The U.S.-Russian space partnership has roots in the ApolloSoyuz Project of the early-to-mid 1970s. The current partnership began in the early 1990s with Phase I (Shuttle-Mir) and continues today with Phase II (ISS) programs. The advantages have been noteworthy: Mir was revitalized and extended because of the capabilities of the Space Shuttle, and Russian assets provided U.S. logistics and astronaut transport to the ISS following the Space Shuttle *Columbia* accident.

In addition to synergies in space exploration, international civil space cooperation has provided positive grounds for fostering understanding between peoples and governments of partner countries. This Committee believes that the existing partnerships should be continued, and it urges the U.S. to consider expanding partnership by forming cooperative relationships with other countries, doing so within the current ISS partnership framework. This framework would serve as the foundation for international collaborations involving exploration beyond low-Earth orbit. This position is consistent with the recent National Research Council report "America's Future in Space: Aligning the Civil Space Program with National Needs," which calls for enhancement of overall U.S. global leadership through global leadership in civil space activities. Specific recommendations address aligning international cooperation with U.S. national interests, which include expanding international cooperation and partnerships for: the study of global climate change; the development of a body of law for a robust space-operating regime; rationalization of export controls; expansion of the ISS partnership; continuation of international cooperation for scientific research and human space exploration; engagement of nations in educating their citizens for sustainable space technology development; and support for interchange among international scholars and students. Partner countries could develop planetary landers, Earth departure stages, habitats, and other significant systems and sub-systems. As the lead partner, the U.S. could develop the heavy-lift launch vehicle. This scenario would help divide costs equitably and elevate the prominence of partner countries.

The Committee notes that China has had an operational human spaceflight program since 2003, and India has announced plans to launch astronauts into space using indigenous assets. The U.S. and India have already begun cooperative activities in space: the Chandrayaan-I spacecraft was flown to the Moon in October 2008 with U.S. instruments onboard. A number of nations are already developing capabilities that could significantly contribute to an international space exploration program. The U.S. has announced that it held preliminary discussions with China regarding joint space activities. It is the view of the Committee that China offers significant potential in a space partnership. China has a human-rated spacecraft and booster system and is only the third country to launch astronauts into space. It has demonstrated advanced capabilities, including extra-vehicular activity on a September 2008 mission. China plans to fly the heavy-lift Long March V vehicle before 2015, which it indicates will eventually be used to establish a space station, currently planned for initial launch in 2020.

8.1.1. International space capabilities

The following summarizes the extent of demonstrated activities in space by nations other than the U.S.

Canada

The Canadian Space Agency (CSA) is a partner in the ISS with significant experience in human operations in space. CSA has specialized in robotics and teleoperated systems, but also has extensive experience in remote sensing, radar, on-orbit servicing, communications and space science.

China

As the third nation in the world to launch humans into space, the People's Republic of China (PRC) has developed an indigenous capability for the technical requirements associated with human spaceflight. The PRC has a stable of space launch vehicles up to medium-lift capability (9 mt to low-Earth orbit, with upgrade to 25 mt envisioned), including the upper stage capability for geo-transfer orbit and interplanetary trajectories. China has launched one robotic probe to the Moon and has follow-on missions planned. The PRC has robust capabilities in communications, navigation and Earth-observation satellites. Next on its announced human spaceflight agenda are demonstrations of rendezvous and docking and the construction of an orbiting space laboratory. The PRC has demonstrated capabilities in life support, power generation and storage, pressurized module construction, in-space propulsion and attitude control, guidance and navigation, communications and computation.

India

The Indian Space Research Organization (ISRO) possesses two very capable launch vehicles, and an upgrade is underway to provide a medium-lift capability. India also has an indigenous capability to produce complex satellites and robotic scientific probes, as demonstrated by their first interplanetary robotic mission to the Moon in 2008. A human spaceflight program is being strongly recommended to the Indian government by ISRO, and is likely to be approved— which could lead to Indian human spaceflight capability as early as 2015. To date, the Indian space program has concentrated on telecommunications, Earth observation, and other low-Earth orbit satellite programs.

Japan

The Japan Aerospace Exploration Agency (JAXA) conducts a robust space program and is a partner in the ISS. Its workhorse launch vehicle, the H-II, has been upgraded to the H-II Transfer Vehicle for use as a logistics carrier to the ISS. While the Japanese space program has a lower launch tempo than other major space-faring nations, it has extensive capabilities, as demonstrated by the ISS Kibo laboratory, which includes teleoperated robotics. Japan has extensive experience in Earth- and space-science missions and telecommunications satellites, as well as in-ground-based facilities for astronaut training, mission operations, communications and tracking. Japan launched the very successful Kaguya lunar robotic mission in 2007 and has plans for follow-on lunar missions.

South Korea

The Republic of Korea has its first space launch vehicle, developed with Russian assistance, ready for flight. Korea has flown as a spaceflight participant to the ISS (via Russian launch and return), but has no other human spaceflight experience. The Republic of Korea has announced plans to develop a lunar orbiter by 2020.

Russia

Russia has a complete suite of space capabilities, from a robust launch vehicle stable to a broad spectrum of spacecraft design, production and operation capabilities. Russia fields a number of space launch vehicles of proven design from small through medium (25 metric tons to low- Earth orbit) with various upper-stage combinations that can provide payloads not only to low-Earth orbit but also to geotransfer orbit and to interplanetary trajectories. Russia is one of three nations to demonstrate the capability to launch humans into space. The highly evolved Soyuz spacecraft is currently programmed to become the linchpin of the ISS in the immediate future. Russia has also demonstrated capabilities in: large space structures; pressurized modules; life support; power generation and storage; communications; thermal control; propulsion and attitude control; guidance and navigation; remote sensing; computation equipment; subsystems; and operations techniques. These are all elements necessary for both human and robotic space exploration. Currently, the Iran, North Korea and Syria Nonproliferation Act (INKSNA) has limited cooperation in space operations between Russia and the United States.

European Space Agency

The European Space Agency (ESA) and its member states possess very significant space capabilities. ESA is a partner in human spaceflight for the ISS and has demonstrated its ability to build large pressurized habitable modules for use as part of the ISS, as well as launch, rendezvous, and other critical capabilities. Through Arianespace (a French company owned by the French government), the Europeans possess the most active commercial space launch program in the world, with various launch vehicles up to medium capability (21 mt to low-Earth orbit). They have demonstrated the capacity to put significant payloads on interplanetary trajectories and have demonstrated space navigation and communications for both low-Earth orbit and interplanetary robotic probes. ESA possesses industrial and commercial capabilities to build complex spacecraft and robotic probes, including all subsystems. The Automated Transfer Vehicle has provided significant logistics support to the

ISS and has the potential to be upgraded to a cargo return vehicle, and eventually a human-carrying spacecraft. Individual member states also have interest in and the capacity for cooperation outside the ESA structure. Cooperation with both ESA and individual European states allows access to significant technological capabilities.

Nations other than the above have very limited space programs, but could potentially play niche roles as their national industrial and technical capabilities allow.

8.2. U.S. Intra-Government Partnerships

The Committee has examined issues about the space programs managed by U.S. government agencies besides NASA, such as the Department of Defense (DoD) and the Commercial Space office of the Federal Aviation Administration (FAA). Their focus includes the potential impacts of "human-rating" the current Evolved Expendable Launch Vehicles (EELV) as one option to reduce the post-Shuttle launch gap. The FAA concentrates on the role commercial space companies can play in the space exploration program.

A study sponsored by the Congress in July 2008 titled "Leadership, Management, and Organization for National Security Space" described the total U.S. space enterprise as including civil space, commercial space, and national security space (military and intelligence focused). All of these elements have overlapping capabilities, share technologies and depend on a common industrial base. As a result, any examination of the U.S. human space program must consider its impact on the efforts of agencies responsible for these other U.S. space activities and *vice versa*.

The most obvious concern has been expressed by the DoD, particularly the U.S. Air Force, because of its unique role today as the single agent for government expendable launch capabilities. The family of Evolved Expendable Launch Vehicles (EELV)—Atlas V and Delta IV—represent the primary heavy-lift capability for the DoD and intelligence communities. Any changes in configuration to these EELV systems, such as incorporating crew escape systems in order to human-rate them, raise concerns about the potential impacts on cost and the ability to reassign vehicles from the EELV production line to high-priority national-security missions.

The Committee reviewed the technical changes required to human rate an EELV. It also met with the National Security Space leadership in the Office of the Secretary of Defense, the Air Force, and the National Reconnaissance Office. Their views on this matter were consistent. None of the national security organizations would formally object to human rating an EELV, if that option were indeed chosen by NASA for its space exploration requirements. However, the concerns noted above were expressed and would need to be managed.

The EELV program was initiated in 1995 in response to the Space Launch Modernization Plan and a subsequent National Space Transportation Policy PDD/NSTC-4. In 1998, development agreements were awarded to Lockheed Martin and Boeing to incorporate government requirements (both civil and national security) into their commercially developed variants of the Atlas and Delta launch systems. Since 2002, the Atlas V and Delta IV EELV vehicles have successfully demonstrated the ability to meet all of the key performance requirements of mass-to-orbit, reliability, and standardization of launch pads and payload interfaces. Because the low-Earth orbit commercial satellite market envisioned in the early-

2000 time period never materialized, the U.S. government is still the primary customer for the EELV systems, which are now operated by United Launch Alliance, a joint venture of Lockheed Martin and Boeing. In 2005, an acquisition strategy change was made to maintain launch and launch-service capabilities for the Atlas and Delta, under a firm, fixed-price EELV launch services contract. This contract provides for a critical mass of sustaining engineering and launch production personnel, even when the launch tempo is low.

The DoD (Air Force) indicates that it is technically feasible to human-rate the EELV systems, as verified for the Committee by an independent Aerospace Corporation study. In doing so, there are several areas that must be addressed. These include:

- Production Facilities – United Launch Alliance is currently consolidating its manufacturing capabilities in Decatur, Alabama. The possibility of human-rating the EELV systems may add complexity to the planning for this consolidation, though the Committee notes that United Launch Alliance has the experience and motivation to mitigate any production conflicts.
- Launch Processing - The Aerospace study commissioned by NASA on Human Rating the Delta IV Heavy informed the Independent Assessment of Launch System Alternatives conducted for the Committee. The study on human rating the Delta IV heavy presented several options to minimize conflict between civil and national security space launch demands. The Aerospace options include the possibility of utilizing the Orbiter Processing Facility and Space Launch Complex 39 at the Kennedy Space Center for processing a first-stage human-rated EELV. Such options need to be further evaluated.
- Cost – The increased production rates stemming from both a human-rated EELV and the national security systems variant of EELV should have a positive effect on United Launch Alliance hardware costs and reliability, as well as on the United Launch Alliance vendor industrial base. A more efficient procurement, surveillance, and mission assurance program should benefit both DoD and NASA programs. Further, the implementation of a human-rated EELV could accelerate the planned transition to a common upper stage for the Delta IV and Atlas V EELVs. Notwithstanding the cost opportunities, the implementation of a human-related EELV does introduce changes in the existing EELV baseline program. Therefore, a comprehensive management plan and structure, with clearly defined responsibilities, authorities and accountabilities, must be formulated if this option is pursued. A similar approach would be required for launch scheduling.
- Industrial Base – If a decision is made to human-rate the EELV systems and NASA were to abandon the Ares I system but retain the Ares V heavy-launch capability, the solid rocket motor industrial base would need to be sustained until the Ares V generated demand. The DoD may have to consider support to the solid rocket motor industrial base in recognition of both civil and NSS needs. If both the Ares I and Ares V programs were abandoned, a detailed civil and military analysis would need to be accomplished to ascertain the interdependence of technical and production capabilities between large solid rocket motors that are needed to support the nation's strategic strike arsenals and the large segmented solid-rocket motors supporting human-rated systems for NASA.

Findings on Partnerships

International partnerships

The U.S. can lead a bold new international effort in the human exploration of space. If international partners are actively engaged, including on the "critical path" to success, there could be substantial benefit to foreign relations, and more overall resources could become available.

National security space

A desirable level of synergy between civil space efforts and national security space efforts should be reached, taking into account the efficient sharing of resources to develop high-value components, as well as the potential challenges in joint management of programs and reliance on a single family of launch vehicles in a class (for example, heavy lift).

9. Concluding Observations

No carefully considered human spaceflight plan, even when promulgated with the best intentions, is likely to produce a successful outcome unless certain principles are embraced in its formulation and execution. Some of the more important of these principles, generally derived from hard-earned experience, are summarized in this final chapter of the report. While not explicitly tasked to offer such observations, the Committee believes that it would be negligent in its duty were it not to do so.

9.1. Establishing Goals

Planning a human spaceflight program should start with agreement about the goals to be accomplished by that program—that is, agreement about its *raison d'être*, not about which object in space to visit. Too often in the past, planning the human spaceflight program has begun with "where" rather than "*why*." This is undoubtedly at least in part attributable to the fact that many of the benefits of human spaceflight are intangible (e.g., the positive impact the Apollo 11 landing had during a time of great tribulation for America). But this makes such intangible benefits and activities no less significant—witness the importance assigned to great literature, music and art in our nation's history.

9.2. Matching Resources and Goals

Perhaps the greatest contributor to risk in the space program, both human and financial, is seeking to accomplish extraordinarily difficult tasks with resources inconsistent with the demands of those tasks. This has undoubtedly been the greatest management challenge faced by NASA in recent decades—even given the magnitude of technological challenges it has confronted. Consider the Constellation Program as a case in point. While it is not clear to the Committee what exactly was the official status of the funding profile NASA assumed in planning the program— there are differing views on the subject—it is clear that the amounts are smaller today by about one-third. It is also clear that when initiating decades-long projects of a demanding technical nature, some baseline funding profile needs to be agreed upon and sustained to the greatest extent practicable.

In the Constellation Program, the estimated cost of the Ares I launch vehicle development increased as NASA determined that the original plan to use the Space Shuttle main engines on the Ares I upper stage would be too costly, in part due to the need to add self-start. But the replacement engine had less thrust and inferior fuel economy, so the first-stage solid rockets had to be modified to provide more total impulse. This in turn contributed to a vibration phenomenon, the correction of which has yet to be fully demonstrated. This is the nature of complex development programs—with budgets that are far more likely to decrease than increase.

Complicating matters further, insofar as the Constellation Program is concerned, this Committee has concluded that the Shuttle Program will almost inevitably extend into FY 2011 in order to fly the existing manifest (the extension largely attributable to safety considerations), and that there are strong arguments for the extension of the International Space Station for another five years beyond the existing plan. These actions, if implemented, place demands of another $1.1 billion and $13.7 billion, respectively, on the NASA budget. In addition, adequate funds must eventually be provided to safely de-orbit the ISS—funds that were not allotted in the current or original program plans.

Shrinking budgets and inadequate reserves—the latter not only in dollars but also in time and technology—are a formula for almost certain failure in human spaceflight. If resources are not available to match established goals, new goals need to be adopted. Simply extending existing ambitious programs "to fit the money" is seldom a solution to the resource dilemma. The impact of fixed costs and technological obsolescence soon overwhelms any such strategy. In the Committee's travels, it encountered widespread support for this policy of realism—although it is likely that most proponents were thinking of having more money, not less program. Should the latter turn out to be the case, much of that conviction is likely to vanish.

In the case of NASA, one result of this dilemma is that in order to pursue major new programs, existing programs have had to be terminated, sometimes prematurely. Thus, the demise of the Space Shuttle and the birth of "the gap." Unless recognized and dealt with, this pattern will continue. When the ISS is eventually retired, will NASA have the capability to pursue exploration beyond low-Earth orbit, or will there be still another gap? When a human-rated heavy-lift vehicle is ready, will lunar systems be available? This is the fundamental conundrum of the NASA budget. Continuation of the prevailing program execution practices (i.e., high fixed cost and high overhead), together with flat budgets, virtually guarantees the creation of additional new gaps in the years ahead. Programs need to be planned, budgeted and executed so that development and operations can proceed in a phased, somewhat overlapping manner.

An additional action that would help alleviate the gap phenomenon is to reenergize NASA's space technology program, an important effort that has significantly atrophied over the years. The role of such a program is to develop advanced components (for example, new liquid rocket engines) that can later be incorporated into major systems. Developing components concurrently with, or as a part of, major system undertakings is a very costly practice. A technology development program closely coordinated with major ongoing programs, but conducted independently of them, is preferable.

9.3. NASA Management Challenges

In planning to reach these lofty objectives with constrained resources, the question arises how NASA might organize to explore. The NASA Administrator, who has been assigned responsibility for the management of NASA, needs to be given the authority to manage NASA. This includes the ability to restructure resources, including workforce and facilities, to meet mission needs. Likewise, managers of programs need clear lines of responsibility and authority. Management of unprecedented and complex international technological developments is particularly challenging, and even the best-managed human spaceflight programs will encounter developmental problems. Such activities must be adequately funded, including reserves to account for the unforeseen and unforeseeable. Good management is especially difficult when funds cannot be moved from one human spaceflight budget line to another, and where new funds can ordinarily be obtained only after a two-year budgetary delay (if at all). In short, NASA should be given the flexibility allowed under the law to acquire and manage its programs.

Fixed overhead and carrying costs at NASA are currently helping to undermine what might be accomplished in new space endeavors. A significant fraction of what appear to be program-related costs in fact cover fixed and carrying costs of employees, facilities and, in some cases, contractors. This reality affects NASA in several ways. When a program such as the Shuttle is terminated, not all of the program funds actually become available to new programs. In fact, the fixed costs often simply move to the new program, where they continue to accumulate. When discretionary resources comprise a limited portion of overall resources, even modest program disruptions can have greatly magnified impacts.

Significant space achievements require continuity of support over many years. One way to assure that *no* successes are achieved is to continually introduce change. Changes to ongoing programs should be made only for compelling reasons. NASA and its human spaceflight program are in need of stability, having been redirected several times in the last decade. On the other hand, decisions about the future should be made by assessing marginal costs and marginal benefits. Sunk costs can never be used as a reason *not* to change. The nation should adopt a long-term strategy for human spaceflight, and changes should be made only for truly compelling reasons. This report describes the advantages and disadvantages of each program option offered. The determination of whether, in balance, these exceed the "compelling reason" threshold is, of course, the essence of making a decision with regard to the future of the human spaceflight program.

There is an often-overlooked but vitally important part of the human spaceflight program that takes place here on Earth. This includes the contributions of the myriad engineers, technicians, scientists and other personnel who work in NASA and industry. As Buzz Aldrin famously said, "It's amazing what one person can do, along with 10,000 friends." Special attention needs to be devoted to assuring the vitality of those portions of the workforce that represent critical and perishable skills that are unique to the space program. One example is the design and manufacturing of very large, solid-propellant motors. At the same time, it is demeaning to NASA's professionalism to treat the human spaceflight effort as a "jobs" program. Only a modest fraction of jobs generally fits the "critical, perishable and unique" criterion. The NASA Administrator needs to be given the authority to tailor the size of the NASA workforce and the number of Centers employing that workforce to match foreseeable needs, much as is routinely done in the private sector under the pressure of competition. For example, when the end of the Cold War changed the role of the aerospace industry, some 640,000 jobs were terminated. Work should be allocated among centers to reflect their legitimate ability to contribute to the tasks to be performed, not simply to maintain a fixed workforce.

NASA's relationship with the private sector requires particularly thoughtful attention. The two entities should not be in competition. NASA is generally at its best when innovating, creating and managing challenging new projects— not when its talents are devoted to more routine functions. Industry is generally at its best when it is developing, constructing and operating systems.

9.4. Systems Engineering

If NASA is to successfully execute the complex undertakings to which it aspires, it must maintain a world-class systems engineering capability, a capability that this and other reviews have deemed to be marginal in its current embodiment. The dilemma is that the best systems engineers are often those with a great deal of experience—"scar tissue," as it is often called by those in the aerospace industry. But how can one get scar tissue if one is confined to studying, analyzing and overseeing the work of others? The answer, by and large, is that one cannot.

One of NASA's answers to this dilemma—which has generated criticism in the past—has been to assume responsibility for developing selected major items of hardware internally (e.g., the Ares I upper stage). This of course places the institution in the hazardous position of serving at once as judge, jury and potential defendant, as well as being in competition with those it manages in other arenas. Thus, NASA finds itself in the position of designing hardware, the engineering drawings for which are being produced by subcontractors to NASA to be handed over to a prime contractor to produce. This sort of formulation warrants exceptionally careful monitoring: it is fraught with opportunities for managerial conflict and technical incompatibility.

A preferred approach for NASA to acquire a strengthened systems engineering capability would be to encourage, or at least permit, the movement of particularly talented individuals back and forth between government and industry, as often occurred during the Apollo Program. This, however, is now discouraged or even precluded by today's government personnel policies (e.g., the long time needed for hiring, well-intentioned but prohibitive conflict of interest policies, etc.). Given this circumstance, the Committee sees no ideal solution for maintaining a strong systems engineering capability at NASA. Perhaps the best among a generally limited array of choices is for NASA occasionally to take direct responsibility for relatively modest pieces of hardware, a responsibility that would not include making or subcontracting engineering or shop drawings for major items to be produced by others. It is noteworthy that the technology development program cited elsewhere in this report could be an effective training ground for systems engineers (as well as program managers), all while maintaining risk at a manageable level.

9.5. Procuring Systems

The Committee has examined various future NASA system options and has observed that in many instances, one of the more significant discriminators in development and operations costs is neither what NASA procures nor who supplies it—but rather how NASA procures and operates a system. The way NASA specifies, acquires, and uses systems; the tools NASA uses to manage its workforce; and the agency's authority to make purchase commitments: all have a very large impact on what NASA can achieve for a given budget.

Currently, NASA labors under many restrictions and practices that impair its ability to make effective use of the nation's industrial base. For example:

- NASA is commonly not allowed to change the size and composition of its workforce or facilities, which limits its ability to save money through the purchase of commercially available products.
- NASA has limited ability to shift funds between related projects to adapt to technical challenges without a protracted approval process.
- NASA is not permitted to make loan guarantees or employ other mechanisms by which it could create a market for commercial providers that might otherwise invest private funds in meeting some of NASA's needs. (The Department of Defense has procurement rules that allow this.) For example, NASA could very likely acquire propellant depots by making a "bankable" commitment to purchase propellant from such a depot; but depending on a "promise" from NASA today would almost certainly not be viewed as a reasonable risk by private investors.
- NASA is expected to undertake long-term projects with little hope of budget stability.

With regard to human spaceflight, it is the Committee's view that NASA can and should be the source of:

- Research and technology
- Technology maturation
- System requirements
- Systems architecture
- Procurement oversight
- Exploration operations
- Expensive, multiple-user facilities

NASA generally should not be its own supplier. Numerous studies have shown that any organization, public or private, that is its own supplier lacks much of the incentive to deliver the most cost-efficient product. Today NASA has many options available to procure systems innovatively. These include (but are not limited to): commercial purchases; Space Act agreements; CCTS-like cost-sharing agreements; prizes for innovative technologies; and others.

Determining the requirements for an engineering project while it is being built inevitably leads to a very expensive result. Requirements should be clearly established prior to beginning engineering development. Work that contains significant risk or for which scope cannot be accurately defined is generally best performed under cost-reimbursable contracts. Work with scope that can be accurately defined should generally be conducted under fixed-price contracts.

The Committee is convinced that NASA can substantially increase the opportunities for entrepreneurial, commercial involvement in its space programs by more aggressively utilizing the commercial authorities already granted to the agency, and by adopting benchmarks in commercial practices utilized in other federal agencies.

9.6. Managing the Balance of Human and Robotic Spaceflight

Although the Committee was tasked only to address the human spaceflight program, including robotic missions that are specifically encompassed within that program, it is appropriate to comment about the role and synergy of human and robotic exploration as a whole. The Committee believes that America is best served by a complementary and balanced space program involving both a robotic component and a human component. The robotic portion is often but not exclusively associated with science missions. Without a strong and sustainable science program—the means of acquiring fundamental new knowledge—any space program would be hollow. The same can be said of the absence of a human spaceflight program. Humans in space, on new and exciting missions, inspire the public. But so do the spectacular accomplishments of such robotic spacecraft as the Hubble Space Telescope, the Mars rovers, the Earth Observing System satellites, or the twin Voyager spacecraft that are poised to reach interstellar space. This is to suggest that both the human spaceflight program and the science program are key parts of a great nation's space portfolio.

Needless to say, robotic spaceflight should play an important role in the human spaceflight program itself, reconnoitering scientifically important destinations, surveying future landing sites, providing logistical support and more. Correspondingly, humans can play an important role in science missions, particularly in field geology, exploration, and the maintenance and enhancement of robotic systems in space. (See Figure 9.6-1.) It is in the interest of both science and human spaceflight that a credible and well-rationalized strategy of coordination between the two types of pursuit be developed—without forcing unwarranted intermingling in areas where each would better proceed on its own.

Robotic activity in space is generally much less costly than human activity and therefore offers a major inherent advantage. Of even greater importance, it does not place human lives at risk. Astronauts provide their greatest advantage in the most complex or novel environments or circumstances. This will be the case in the exploration of planetary surfaces and in repair or servicing missions of the type undertaken for the Hubble Space Telescope's primary mirror. In contrast, the value of humans in space is usually at its minimum when they are employed transporting cargo. The bottom line is that there are important roles to be played by both humans and robots in space, and America should strive to maintain a balanced program incorporating the best of both kinds of explorers.

That said, there are nonetheless inevitable conflicts— conflicts that arise from the competition among programs for resources, particularly financial resources. It is therefore of the utmost importance, if balance is to be maintained, that neither the human program nor the robotic spaceflight program be permitted to cannibalize the other. This has been a significant concern in the past, particularly given the size of the human spaceflight program. Difficulties in the human space program too often swallowed resources that had been planned for the robotic program (as well as for aeronautics and space technology). Robotics are generally, although not exclusively, considered to be of greater interest to the scientific community. It is essential that *budgetary* firewalls be built between these two broad categories of activity. In the case of the International Space Station, one firewall should be the establishment of an organizational entity to select endeavors to be pursued aboard the Space Station. Without such a mechanism, turmoil is assured and program balance endangered.

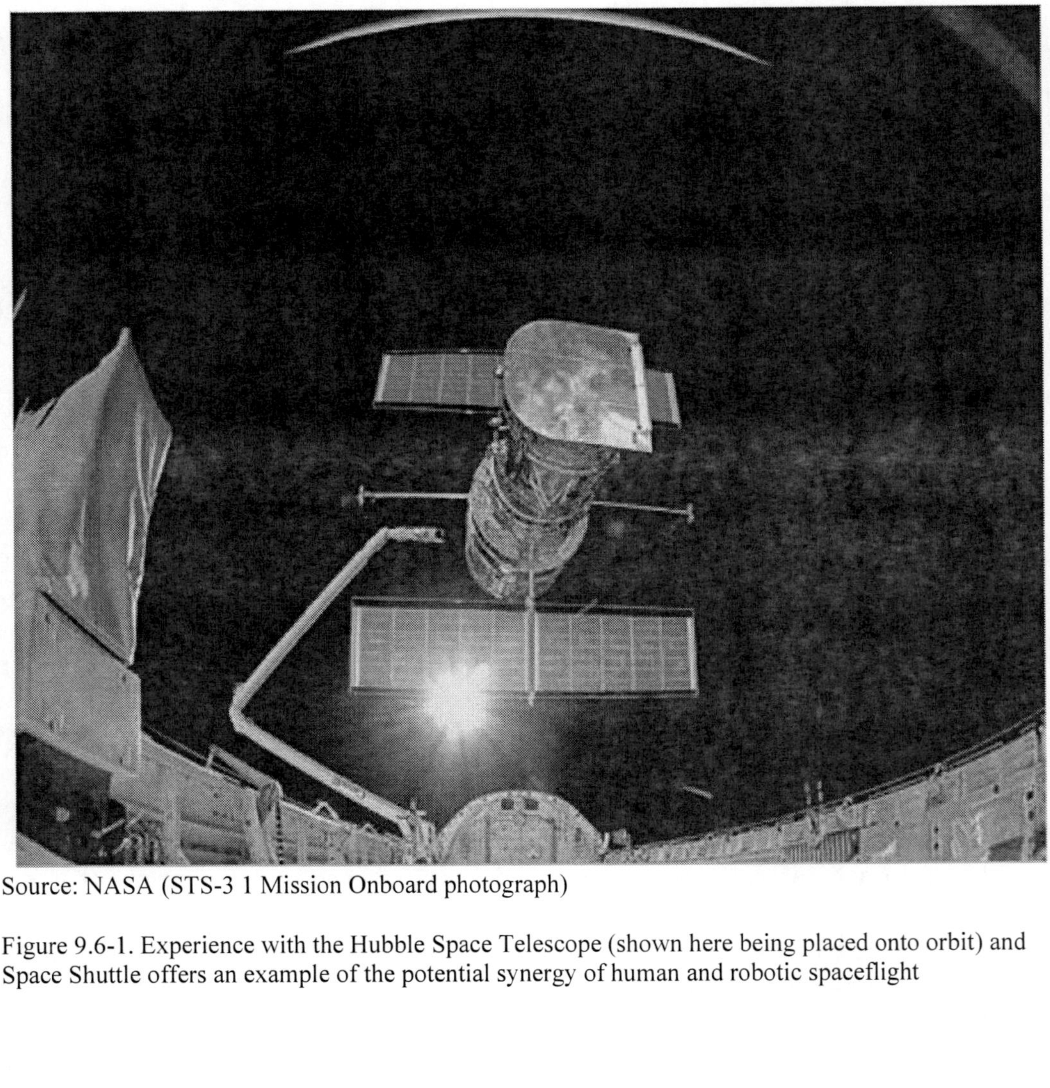

Source: NASA (STS-3 1 Mission Onboard photograph)

Figure 9.6-1. Experience with the Hubble Space Telescope (shown here being placed onto orbit) and Space Shuttle offers an example of the potential synergy of human and robotic spaceflight

Findings

The right mission and the right size

NASA's budget should match its mission and goals. Further, NASA should be given the ability to shape its organization and infrastructure accordingly, while maintaining facilities deemed to be of national importance.

Robotic program coordination

The robotic and human explorations of space should be synergistic, both at the program level (e.g., science probes to Mars and humans to Mars) and at the operational level (e.g., humans with robotic assistants on a spacewalk). Without burdening the space science budget or influencing its process of peer-based selection of science missions, NASA should proceed to develop the robotic component of its human exploration program.

Management authority

The NASA Administrator and program managers need to be given the responsibility and authority to manage their endeavors. This includes providing flexibility to tailor resources, including people, facilities and funds, to fit mission needs.

Stability in programs

In the most recent decade, NASA has spent about 80 percent of the GDP-deflated budget that it had in the decade of Apollo. Recurring budget ambiguities and reductions and redirections of policy, coupled with the high fixed-cost structure of NASA, have not optimized the return on that investment.

Right job for the NASA workforce

NASA has a talented (but aging) workforce. NASA should focus on the challenging, long-term tasks of technology development, cutting-edge new concepts, system architecture development, requirements definition, and oversight of the development and operation of systems.

Fixed operating costs at NASA

There are significant fixed costs in the NASA system. Given that reality, reducing the funding profile much below the optimum for the development of a given program has an amplified effect of delaying benefits and increasing total program cost.

NASA's fundamental budgetary conundrum

Within the current structure of the budget, NASA essentially has the resources either to build a major new system or to operate one, but not to do both. This is the root cause of the gap in capability of launching crew to low-Earth orbit under the current budget and will likely be the source of other gaps in the future.

Commercial involvement in exploration

NASA has considerable flexibility in its acquisition activities due to special provisions of the Space Act. NASA should exploit these provisions whenever appropriate, and in general encourage more engagement by commercial providers, allocating to them tasks and responsibilities that are consistent with their strengths.

9.7. Concluding Summary

NASA is the most accomplished space organization in the world. Its human spaceflight activities are nonetheless at a tipping point, primarily due to a mismatch of goals and resources. Either additional funds need to be made available or a far more modest program involving little or no exploration needs to be adopted. Various options can be identified that offer exciting and worthwhile opportunities for the human exploration of space if appropriate funds can be made available. Such funds can be considerably leveraged by having NASA attack its overhead costs and change some of its traditional ways of conducting its affairs—and by giving its management the authority to bring about such changes. The American public can take pride in NASA's past accomplishments; the opportunity now exists to provide for the future human spaceflight program worthy of a great nation.

Appendix A. Committee Member Biographies

Norman R. Augustine

Norman R. Augustine graduated from Princeton University where he was awarded Bachelor's and Master's degrees in aeronautical engineering and was elected to Phi Beta Kappa, Tau Beta Pi and Sigma Xi. He is a retired Chairman and CEO of Lockheed Martin Corporation and has served as Assistant Secretary, Under Secretary and Acting Secretary of the Army and as an Assistant Director of Defense Research and Engineering. He has served as Chairman of the National Academy of Engineering, the Defense Science Board and the Aerospace Industries Association, and as President of the American Institute of Aeronautics and Astronautics. A 16-year member of the President's Council of Advisors on Science and Technology, Mr. Augustine chaired the Aeronautics Committee of the NASA Advisory Council and served on the Air Force Scientific Advisory Board. He is an Honorary Fellow of the American Institute of Aeronautics and Astronautics and a Fellow of the American Society of Mechanical Engineers, the Institute of Electrical and Electronic Engineers and the International Academy of Astronautics. A former member of the faculty of the Princeton University Department of Mechanical and Aerospace Engineering, he has received the National Medal of Technology, awarded by the President of the United States, and holds 23 honorary degrees.

Chairman

Wanda M. Austin

Wanda M. Austin earned a Bachelor's Degree in mathematics from Franklin & Marshall College, Master's degrees in systems engineering and mathematics from the University of Pittsburgh, and a Ph.D. in systems engineering from the University of Southern California. She is President and CEO of The Aerospace Corporation, a leading architect for the nation's national security space programs. She is a member of the National Academy of Engineering, a Fellow of the American Institute of Aeronautics and Astronautics, and a member of the International Academy of Astronautics. She was a member of the National Research Council's Committee on the Rationale and Goals of the U.S. Civil Space Program. She previously served on the NASA Advisory Council and the NASA Aerospace Safety Advisory Panel. Dr. Austin has received the National Intelligence Medallion for Meritorious Service, the Air Force Scroll of Achievement, the National Reconnaissance Office Gold Medal, the U.S. Air Force Meritorious Civilian Service Medal and the NASA Pubic Service Medal. She was inducted into the "Women in Technology International" Hall of Fame, and she has been named the 2009 Black Engineer of the Year. Dr. Austin is internationally recognized for her work in satellite and payload system acquisition, systems engineering and system simulation.

Bohdan I. Bejmuk

Bohdan (Bo) I. Bejmuk is an aerospace consultant with in-depth knowledge of space systems and launch vehicles. He is Chairman of the Standing Review Board for the NASA Constellation

Program. Mr. Bejmuk retired from the Boeing Company in 2006, where he was the Space Shuttle Orbiter Program Director, responsible for all Orbiter engineering efforts. During Space Shuttle development and early operations, he served as Program Manager for system engineering and integration. On Sea Launch, an international joint venture, he was the Executive Vice President and Chief Engineer, directing all aspects of the company's development and leading a multinational workforce of several thousand engineers and shipyard workers. After completion of Sea Launch development, he was manager of operations at the California home port and at the Pacific Ocean launch region. Mr. Bejmuk also served in numerous senior positions at Rockwell International and Martin Marietta. He received a B .S. and M.S. in mechanical engineering from the University of Colorado. A member of the International Academy of Astronautics, he received the Lloyd V. Berkner Award, the Aviation Week Laurels Award, the National Public Service Medal, awarded twice by NASA, and the Rockwell International Presidents Award. He was also recognized as an Eminent Engineer by the California Institute of Technology.

Leroy Chiao

In 15 years with NASA, astronaut Leroy Chiao logged more than 229 days in space, including 36 hours in extra-vehicular activity. A veteran of four space missions, Dr. Chiao most recently served as Commander and NASA Science Officer of Expedition 10 aboard the International Space Station. He was a Space Shuttle Mission Specialist and also certified as a copilot of the Russian Soyuz spacecraft. Since leaving NASA in 2005, Dr. Chiao has worked with entrepreneurial business ventures in the U.S., China, Japan and Russia. Among these, he is Executive Vice President and a Director of Excalibur Almaz, a private manned spaceflight company. He serves as Chairman of the National Space Biomedical Research Institute User Panel and advisor and spokesman for the Heinlein Prize Trust. Dr. Chiao is a Director of the Challenger Center for Space Science Education and of the Committee of 100, an organization of prominent U.S. citizens of Chinese descent. Dr. Chiao earned a B .S. from the University of California at Berkeley and his M.S. and Ph.D. at the University of California at Santa Barbara, all in chemical engineering. Prior to joining NASA, he worked as a research engineer at Hexcel Corp. and then at the Lawrence Livermore National Laboratory.

Christopher Chyba

Christopher Chyba is Professor of Astrophysical Sciences and International Affairs at Princeton University. He previously was Associate Professor of Geological and Environmental Sciences at Stanford University, and held the Carl Sagan Chair at the SETI

Institute. Dr. Chyba served at the White House from 1993 to 1995, entering as a White House Fellow on the National Security Council staff, and then in the Office of Science and Technology Policy. In 1996, Dr. Chyba received the Presidential Early Career Award for Scientists and Engineers. In 2001, he was named a MacArthur Fellow for his work in both international security and planetary science. Dr. Chyba serves on the National Academy of Sciences Committee on International Security and Arms Control and is past Chair of the National Research Council Committee on Preventing the Forward Contamination of Mars. He has served on NASA's Space Science Advisory Committee, for which he chaired the Solar System Exploration Subcommittee, and he chaired the Science Definition Team for NASA's Europa Orbiter mission. A physics graduate of Swarthmore College, Dr. Chyba holds an M.Phil. from Cambridge University, where he was a Marshall Scholar, and a Ph.D. in astronomy and space sciences from Cornell University. In 2009, President Obama appointed him to the President's Council of Advisors on Science and Technology.

Edward F. Crawley

Edward F. Crawley is the Ford Professor of Engineering at MIT, and is a Professor of Aeronautics and Astronautics and of Engineering Systems. He received an S .B. and Sc.D. in aerospace engineering from MIT, and he holds an honorary Doctorate from Chalmers University. He has served as the Head of the MIT Department of Aeronautics and Astronautics, the Director of the Cambridge – MIT Institute, and the Director of the Bernard M. Gordon – MIT Engineering Leadership Program. His research has focused on the domain of architecture, design and decision support in complex technical systems. Dr. Crawley is a Fellow of the AIAA and the Royal Aeronautical Society (UK), and is a member of three national academies of engineering: in Sweden, the UK and the United States. Dr. Crawley served as Chairman of the NASA Technology and Commercialization Advisory Committee, and he was a member of the 1993 Presidential Advisory Committee on the Space Station Redesign. He recently co-chaired the committee of the NRC reviewing the NASA Exploration Technology Development Program. He served as a lecturer at the Moscow Aviation Institute, and as a Guest Professor at Tsinghua University in Beijing. In 1980 he was a finalist in the NASA astronaut selection. He has founded three entrepreneurial companies and currently sits on several corporate boards.

Jeff Greason

Jeff Greason, CEO of XCOR Aerospace, has 18 years of experience managing innovative, leading- edge technical project teams at XCOR Aerospace, Rotary Rocket and Intel Corporation. At XCOR, he leads a team that designs, builds and operates long-life, low-cost, reusable rocket engines and rocket- powered vehicles for government and private markets. During his work at XCOR, Mr. Greason has had final go/no-go authority on more than 20 manned rocket flights and hundreds of rocket engine tests. The company has won and successfully completed government contracts for NASA, the U.S . Air Force and DARPA. A recognized expert in reusable launch vehicle regulations of the Federal Aviation Administration's Office of Commercial Space Transportation, he testified before the joint House/Senate subcommittee hearings on Commercial Human Spaceflight, which led to the Commercial Space Launch Amendments Act of 2004. He serves on the Commercial Space Transportation Advisory Committee (COMSTAC) and is a co-founder and Vice Chairman of the Personal Spaceflight Federation, a trade association for innovative launch companies. Mr. Greason was cited by Time magazine in 2001 as one of the "Inventors of the Year" for his team's work on the EZ-Rocket. At Intel, he developed leading-edge processor design techniques and received the coveted Intel Achievement Award. He holds 18 U.S. patents and graduated with honors from the California Institute of Technology.

Charles F. Kennel

Charles F. Kennel earned his A.B. at Harvard University and his Ph.D at Princeton University, studying space plasma physics and astrophysics. After three years at the Avco-Everett Research Laboratory, he joined the UCLA Physics Department, eventually chairing the department, and TRW Systems. He served as: NASA Associate Administrator for Mission to Planet Earth; UCLA Executive Vice Chancellor; Director of the Scripps Institution of Oceanography; and Vice Chancellor of Marine Sciences, University of California, San Diego. Dr. Kennel is a member of the National Academy of Sciences, the American Academy of Arts and Sciences, the American Philosophical Society and the International Academy of Astronautics. He has chaired the National Academy's Solar and Space Physics, Global Change Research, Fusion Sciences, and Beyond Einstein committees, as well as its Board on Physics and Astronomy. He was a member of the Pew Oceans Commission and both member and Chair of the NASA Advisory Council. He now chairs the California Council on Science and Technology and the NRC Space Studies Board, and is a Board member of the Bermuda Institute of Ocean Sciences. Dr. Kennel has received prizes from the American Physical Society, the European Geophysical Union and the Italian Academy, both the NASA Distinguished Service and Distinguished Public Service medals, and an honorary degree from the University of Alabama.

Lester L. Lyles

Lester L. Lyles, a retired U.S. Air Force four-star general, graduated from Howard University with a Bachelor's Degree in mechanical engineering. He began his 35-year career in the Air Force as a space vehicle engineer, after earning a Masters Degree in mechanical/nuclear engineering from New Mexico State University. Following the Shuttle Challenger accident, General Lyles directed the recovery operations conducted by the Air Force Space-Launch Systems Office. For this effort, the National Space Club recognized him as the Astronautics Engineer of the Year. General Lyles commanded the Ogden Air Logistics Center and the Space and Missile Systems Center, and he directed the Ballistic Missile Defense Organization. He served as the 27th Vice Chief of Staff of the U.S. Air Force and Commander of the Air Force Materiel Command. He served as a member of the President's Commission on the Implementation of the U.S. Space Exploration Vision, and he chaired the National Research Council study on the Rationale and Goals of the U.S. Civil Space Program. He chairs the Aeronautics Committee of the NASA Advisory Council, and he is the Vice Chair of the Defense Science Board. His numerous honors include the Black Engineer of the Year Lifetime Achievement Award, as well as two honorary doctoral degrees.

Sally K. Ride

Sally K. Ride earned a B.S. in physics and B.A. in English, followed by her M.S. and Ph.D. in physics, all at Stanford University. She is a Professor of Physics (Emerita) at the University of California, San Diego, and the CEO of Imaginary Lines, Inc. Dr. Ride, the first American woman in space, was an astronaut for more than 10 years, flying on two Space Shuttle missions. She was NASA's first Director of Exploration and first Director of Strategic Planning, and she is the only person to serve on both the Space Shuttle Challenger and Columbia accident investigation boards. A Fellow of the American Physical Society and Board Member of both the California Institute of Technology and the Aerospace Corporation, she has also served on the Space Studies Board, the Board of the Congressional Office of Technology Assessment, the President's Commission on White House Fellows and the President's Committee of Advisors on Science and Technology. Dr. Ride has received the Jefferson Award for Public Service, the Von Braun Award and the Lindbergh Eagle, and she has twice been awarded the National Spaceflight Medal. She is a member of the Aviation Hall of Fame and the Astronaut Hall of Fame.

Appendix B. Committee Staff

Designated Federal Official/Executive Director

Philip McAlister, Special Assistant for Program Analysis,
NASA Headquarters

NASA

NASA Chief of Staff

George Whitesides, NASA Headquarters

NASA Review Team

Study Team

Dr. W. Michael Hawes, HQ - Team Lead

Brent Jett, JSC - Deputy

Tricia K. Mack, HQ - Executive Officer

Program Risk Assessment Team

David Radzanowski, HQ - Team Lead

Frank H. Bauer, HQ

Glenn Butts, KSC

Melek Ferrara, SAIC

Eameal Holstien, JSC

Charlie Hunt, HQ

Mahmoud R. Naderi, MSFC

Josh Manning, HQ

Stanley McCaulley, MSFC

Carey McCleskey, KSC

Arlene Moore, HQ

Krista Paquin, HQ

Tom Parkey, GRC

Frank A. (Andy) Prince, MSFC

Patrick Scheuermann, SSC

Daniel Schumacher, MSFC

Robert Sefcik, GRC

Christian Smart, Ph.D., SAIC

Edgar Zapata, KSC

Engineering Analysis

Ralph Roe, HQ - Team Lead

Mike Kirsch, LaRC - Deputy

Dawn Schaible, LaRC – Deputy

Reginald Alexander, MSFC

John D. Baker, JPL

Keith Brock, SSC

William M. Cirillo, LaRC

Scott Colloredo, KSC John Connolly, JSC

Terri Derby, ATK Space Systems

Bret Drake, JSC

Dennis Hines, DFRC
Eric Isaac, GSFC
Michael A. Johnson, GSFC
Kent Joosten, JSC
Dave Korsmeyer, ARC
Steven Labbe, JSC
Rick Manella, GRC
Roland Martinez, JSC
Paul McConnaughey, MSFC
Pat Troutman, LaRC

Engineering Analysis Support

Dan Adamo (ret'd), JSC
Rob Adams, MSFC
Faye Allen, Futron Corp.
David Anderson, MSFC
Mark Andraschko, LaRC
Michael Baine, JSC
Stanley Borowski, GRC
Robert Bruce, SSC
Sue Burns, JSC
Robert Cataldo, GRC
Grant R. Cates, Ph.D., SAIC
David Chato, GRC
Robert Christy, GRC
Kristine Collins, LaRC
John Cornwell, JSC
Nicholas Cummings, KSC
Dominic DePasquale, SEI
Ian Dux, GRC
Kevin Earle, LaRC
Walt Engelund, LaRC
Robert Falck, GRC
James Fincannon, GRC
Brad Flick, DFRC
Joseph R. Fragola, Valador Inc
Joseph Gaby, ASRC Ricardo Galan, SAIC
Dr. David M Garza, Aerospace
Leon Gefert, GRC
James Geffre, JSC
Justin Gelito, SAIC
Patrick George, GRC
John C. Goble, Aerospace
Reynaldo Gomez, JSC
Kandyce Goodliff, LaRC
Joe Grady, GRC

Brand Griffin, Gray Research
John Gruener, JSC
Gene Grush, JSC
Kurt Hack, GRC
Leon Hastings, MSFC
Edward Henderson, JSC
Stephen Hoffman, SAIC
Dr. Robert L. Howard, Jr., JSC
A. Scott Howe, JPL
Eric Hurlbert, JSC
Daniel C. Judnick, Aerospace
Kriss Kennedy, JSC
D.R. Komar, LaRC
Larry Kos, MSFC
Rob Landis, ARC
Jeffrey A. Lang, Aerospace
Roger Lepsch, LaRC
Dr. Marcus A. Lobbia, Aerospace
Gaspare Maggio, ISL
Richard Mattingly, JPL
Nicholas E. Martin, Aerospace
Daniel Mazanek, LaRC
Dave McCurdy, ARSC
Melissa L. McGuire, GRC
Pat McRight, MSFC
Raymond (Gabe) Merrill, LaRC
Michael Meyer, GRC
Glenn Miller, JSC
Dr. Inki A. Min, Aerospace
Nathaniel Morgan, LaRC
Jack Mulqueen, MSFC
Tri X. Nguyen, JSC
Steven Oleson, GRC
Tom Packard, Analex Corp.
Bill Pannell, MSFC
Robert Phillips, Futron Corp.
Charles Pierce, MSFC
Blake F. Putney Jr., Valador, Inc.
Shawn Quinn, KSC
David Reeves, LaRC
J.D. Reeves, LaRC
Joe Roche, GRC
Marianne Rudisill, LaRC
Joshua Sams, AMA, Inc.
Rudolph Saucillo, LaRC
Larry Schultz, KSC

Hilary Shyface, AMA, Inc.
Gary Spexarth, JSC
Jonette Stecklein, JSC
Chel Stromgren, SAIC
Phil Sumrall, MSFC
Ted Sweetser, JPL
Ted Talay, JF&A
William Taylor, GRC
Chris Teixeira, SAIC
Andrew Thomas, JSC
Ed Threet, MSFC
Larry Toups, JSC
Stephen Tucker, MSFC
Eugene Ungar,
JSC Timothy Verhey, GRC
Phil Weber, KSC
Robert Werka, MSFC
Charles Whetsel,
JPL Doug Whitehead, JSC
Jeff Woytach, GRC

Strategic Analysis and Collaboration

Tom Cremins, HQ - Team Lead
Dave Bartine, KSC
Dennis Boccippio, MSFC
John Casani, JPL
Doug Comstock, HQ
Jason Derleth, HQ
Karen Feldstein, HQ
Chris Ferguson, JSC
Lee Graham, JSC
Richard B. Leshner, HQ
Neal Newman, HQ
Mark Uhran, HQ

Science and Technology

Michael Meyer, HQ – Team Lead
Ariel Anbar, AZ State University
Dave Beaty, JPL
Francis A. Cucinotta, JSC
Jeffrey R. Davis, JSC
James Head, Brown University
Terri Lomax, NC State University

John Mustard, Brown University
Clive Neal, Univ. of Notre Dame
Edward Semones, JSC
Mark Sykes, Planetary Sci. Inst.
Meenakshi Wadhwa, AZ State Univ.

Programs, Legislative Affairs, Legal Counsel, Public Affairs, External Affairs

Susan Burch, HQ
Rebecca Gilchrist, HQ
Bill Hill, HQ
Margaret Kieffer, HQ
Mary D. Kerwin, HQ
Doc Mirelson, HQ
Diane Rausch, HQ
Kirk Shireman, JSC
Kathleen Teale, HQ

Astronaut Representatives

Chris Ferguson, JSC
Don Pettit, JSC
Andy Thomas, JSC

NASA Television

Fred A. Brown, Team Lead
Allard Beutel, KSC
Cliff Feldman
Mark Hailey
Cedric Harris
Bill Hubscher, MSFC
John Malechek
Bob Moder, MSFC
Anthony Stewart
Jennifer Tharpe,
KSC Robert Williams

Committee Site Visit Support

Sandra H. Turner MSFC
Patsy H. Fuller, MSFC

Valador, Inc. Support

Donna Connell, Team Lead
Dennis R. Bonilla
Lisa S. Connell
James Convy IV

Kathleen Gallagher
Sandra L. George
Matthew Kohut
Erek Korgan
Stephen F. Spiker
Kelly E. Stocks
Lester A. Reingold
Tate S. Srey
Patricia D. Trenner
Joseph A. Valinotti III

Executive Assistant to Norman R. Augustine

Laura J. Ahlberg

Aerospace Corporation Support

Gary P. Pulliam - Vice President
Matthew J. Hart - Team Lead
John P. Skratt - Team Lead
Dr. Inki A. Min - Team Lead
Debra L. Emmons - Team Lead
Dr. David A. Bearden - Team Lead
David S. Adlis Joseph A. Aguilar
Dr. Kristina Alemany
Violet Barghe-Sharghi
Stephanie E. Barr
Robert E. Bitten
Garry H. Boggan
John J. Bohner
Daniel W. Bursch
Vincent M. Canales
Dr. Shirin Eftekhardezeh
Debra L. Emmons
Dr. Robert W. Francis
Jonathan P. French
Gregory Fruth
Larcine Gantner
Alisa M. Hawkins
Thanh D. Hoang
Thomas P. Jasin
Ray F. Johnson
Daniel C. Judnick
Randolph L. Kendall
Dr. Robert J. Kinsey
Jeffrey A. Lang
Dr. Tung T. Lam

Glenn W. Law
Dr. Austin S. Lee
Jeffrey R. Lince
Dr. Marcus A. Lobbia
Nicholas E. Martin
John P. Mayberry
Shannon P. McCall
Michael R. Moore
Elisha B. Murrell
Dr. Jon M. Neff
Daniel A. Nigg
Thomas Paige
Jay P. Penn
Dr. Torrey O. Radcliffe
Ernest Y. Robinson
Louis H. Sacks
Felix T. Sasso
Maria E. Sklar
Morgan W. Tam
Paul R. Thompson
E. J. Tomei, Jr.
Dr. Leslie A. Wickman
Justin S. Yoshida
Dr. Albert H. Zimmerman

APPENDIX C. REVIEW OF U.S. HUMAN SPACE FLIGHT PLANS STATEMENT OF TASK

This Statement of Task establishes and informs a review to be conducted in support of planning for U.S. human space flight activities beyond the retirement of the Space Shuttle. The purpose of this effort is to develop suitable options for consideration by the Administration regarding a human space flight architecture that would:

- Expedite a new U.S. capability to support utilization of the International Space Station
- Support missions to the Moon and other destinations beyond low Earth orbit (LEO)
- Stimulate commercial space flight capability
- Fit within the current budget profile for NASA exploration activities

The review will be led by an independent, blue-ribbon panel of experts who will work closely with a NASA team and will report progress on a regular basis to NASA leadership and the Executive Office of the President. This independent review will provide options and related information to involved Administration agencies and offices in sufficient time to support an August 2009 decision on the way forward. As necessary and appropriate, the team may seek early decisions from the Administration on some of these options. A final report containing the options and supporting analyses from this review also will be released.

Scope

The review should:

- Evaluate the status and capabilities of the agency's current human space flight development program;
- Evaluate other potential architectures that are capable of supporting the mission areas described above;
- Evaluate what capabilities and mission scenarios would be enabled by the potential architectures under consideration, including various destinations of value beyond LEO;
- Consider options to extend International Space Station operations beyond 2016;
- Examine the appropriate degree of R&D and complementary robotic activities necessary to make human space flight activities affordable and productive over the long term;
- Examine appropriate opportunities for international collaboration; and
- Not rely upon extending Space Shuttle operations in assessing potential architectures.

The review may evaluate architectures that build on current plans, existing launch vehicles and infrastructure, Space Shuttle- related components and infrastructure, the two Evolved Expendable Launch Vehicle (EELV) families, and emerging capabilities. It may also consider architectures that vary in terms of the capability that would be delivered beyond low Earth orbit (e.g., the number of crew and the duration of these missions), while describing the implications of such choices for possible mission goals and scenarios. In addition to new analyses required in support of this effort, the review team should consider, where appropriate, other studies and reports relating to this subject.

Evaluation Parameters

The review should examine potential architectures relative to the following key evaluation parameters:

- Crew (and overall mission) safety;
- Overall architecture capability (e.g., mission duration, mass delivered to low Earth orbit and other selected destinations, flexibility);
- Life-cycle costs (including operations costs) through 2020;
- Development time;
- Programmatic and technical risk;
- Potential to spur innovation, encourage competition, and lower the cost of space transportation operations in the existing and emerging aerospace industry;

- Implications for transition from current human space flight operations;
- Impact on the nation's industrial base and competitiveness internationally;
- Potentially expanded opportunities for science;
- Potential for enhanced international cooperation as appropriate;
- Potential to enhance sustainability of human space activities;
- Potential for inspiring the nation, and motivating young people to pursue careers in science, technology, engineering and mathematics subjects;
- Benefit to U.S. Government defense and intelligence space-related capabilities; and
- Contractual implications.

Budget

Budget options considered under the review must address the development of a human space flight architecture, robotic spacecraft to support and complement human activities, and R&D to support future activities. The review should assume the following 2010-2014 budget profile for these activities:

2010	2011	2012	2013	2014
3,963.1	6,092.9	6,077.4	6,047.7	6,274.6

($ in millions)

Based on the results of this review, the Administration will notify Congress of any needed changes to the FY20 10 President's Budget Request.

APPENDIX D. CHARTER OF THE REVIEW OF U.S. HUMAN SPACE FLIGHT PLANS COMMITTEE

1. **Official Designation:** Review of U.S. Human Space Flight Plans Committee ("The Committee")

2. **Authority:** Having determined that it is in the public interest in connection with the performance of Agency duties under law, and in consultation with the U.S. General Services Administration, the NASA Administrator hereby establishes the Review of U.S. Human Space Flight Plans Committee pursuant to the Federal Advisory Committee Act (FACA), as amended, 5 U.S.C. App.

3. **Scope and Objectives:** The Committee shall conduct an independent review of ongoing U.S. human space flight plans and programs, as well as alternatives, to ensure the nation is pursuing the best trajectory for the future of human space flight – one that is safe, innovative, affordable, and sustainable. The Committee should aim to identify and characterize a range of options that spans the reasonable possibilities

for continuation of U.S. human space flight activities beyond retirement of the Space Shuttle. The identification and characterization of these options should address the following objectives: a) expediting a new U.S. capability to support utilization of the International Space Station (ISS); b) supporting missions to the Moon and other destinations beyond low Earth orbit (LEO); c) stimulating commercial space flight capability; and d) fitting within the current budget profile for NASA exploration activities.

In addition to the objectives described above, the review should examine the appropriate amount of R&D and complementary robotic activities needed to make human space flight activities most productive and affordable over the long term, as well as appropriate opportunities for international collaboration. It should also evaluate what capabilities would be enabled by each of the potential architectures considered. It should evaluate options for extending International Space Station operations beyond 2016.

4. **Description of Duties:** The Committee will provide advice only.

5. **Official to Whom the Committee Reports:** The Committee reports to the NASA Administrator and the Director of the Office of Science and Technology Policy (OSTP), Executive Office of the President. The Committee will submit its report within 120 days of the first meeting of the Committee.

6. **Support:** The NASA Office of Program Analysis and Evaluation shall provide staff support and operating funds for the Committee.

7. **Estimated Annual Operating Costs and Staff Years:** The operating cost associated with supporting the Committee's functions is estimated to be approximately $3 million, including all direct and indirect expenses. It is estimated that approximately 8 full-time equivalents will be required to support the Committee.

8. **Designated Federal Officer:** The Executive Director of the Committee shall be appointed by the NASA Administrator and shall serve as the Designated Federal Official (DFO). The DFO must be either a full-time or a permanent part-time employee, who must call, attend, and adjourn committee meetings; approve agendas; maintain required records on costs and membership; ensure efficient operations; maintain records for availability to the public; and provide copies of committee reports to the NASA Committee Management Officer (CMO) for forwarding to the Congress.

9. **Estimated Number and Frequency of Meetings:** The Committee shall conduct meetings as appropriate at various locations throughout the United States. Meetings shall be open to the public unless it is determined that the meeting, or a portion of the meeting, will be closed in accordance with the Government in the Sunshine Act.

10. **Duration:** The Committee will exist for 180 days, unless earlier renewed.

11. **Termination:** The Committee shall terminate within 60 days after submitting its report.

12. **Membership and Designation:** The Committee shall consist of members to be appointed by the NASA Administrator. The Administrator shall ensure a balanced representation in terms of the points of view represented and the functions to be performed. Each member serves at the pleasure of the Administrator. The Committee shall consist of approximately 5-10 members. It is anticipated that the members will serve as Special Government Employees for the duration of the Committee, renewable at the discretion of the NASA Administrator. The NASA Administrator shall designate the chair of the Committee.

13. **Subcommittees:** Subcommittees, task forces, and/or work groups may be established by NASA to conduct studies and/or fact-finding requiring an effort of limited duration. Such subcommittees, task forces and work groups will report their findings and recommendations directly to the Committee. However, if the Committee is terminated, all subcommittees, task forces and work groups will also terminate.

14. **Record keeping:** The records of the Committee, formally and informally established subcommittees, or other subgroups of the Committee, shall be handled in accordance with General Records Schedule 26, Item 2, or other approved agency records disposition schedule. These records shall be available for public inspection and copying, subject to the Freedom of Information Act, 5 U.S.C. 552.

15. **Charter Filing Date:** This charter shall become effective upon the filing of this charter with the appropriate U.S. Senate and House of Representatives oversight committees.

June 1, 2009

Christopher J. Scolese
NASA Administrator (Acting)

Date

Appendix E. List of Full Committee Meetings and Locations

Date	Meeting
June 9, 2009	Preparatory Meeting (Teleconference)
June 16, 2009	Preparatory Meeting (Washington, D.C.)
June 17, 2009	Public Meeting (Washington, D.C.)
June 18, 2009	Site Visit (Dulles, VA)
June 24-25, 2009	Site Visit (Huntsville and Decatur, AL; and Michoud, LA)
July 2, 2009	Preparatory Meeting (Teleconference)
July 8-9, 2009	Site Visit (Hawthorne, CA) and Fact-Finding Meetings (El Segundo, CA)

July 14, 2009 Preparatory Meeting (Teleconference)
July 21-23, 2009 Fact-Finding Meetings (Denver, CO)
July 28, 2009 Public Meeting (Houston, TX)
July 29, 2009 Public Meeting (Huntsville, AL)
July 30, 2009 Public Meeting (Cocoa Beach, FL)
August 5, 2009 Public Meeting (Washington, D.C.)
August 5, 2009 Preparatory Meeting (Washington, D.C.)
August 12, 2009 Preparatory Meeting (Washington, D.C.)
August 12, 2009 Public Meeting (Washington, D.C.)
October 8, 2009 Public Meeting (Teleconference)

APPENDIX F. BRIEFERS AND COMMITTEE CONTACTS

The following is a list of individuals who briefed the Committee or responded to its requests for information:

E.C. "Pete" Aldridge
Buzz Aldrin
Brett Alexander
Reginald Alexander
John D. Baker
Frank H. Bauer
Jeanne L. Becker
James M. Beggs
Dallas Bienhoff
Jack Bullman
Jack O. Burns
Frank Buzzard
Bob Cabana
Elizabeth Cantwell
Frank Chandler
Jim Chilton
Lynn Cline
Mike Coats
Cassie Conley
Doug Cooke
Ed Cortwright
Dick Covey
William M. Cirillo
Steve Creech
Chris Culbert
Danny Davis
Jean-Jacques Dordain
Bret Drake
Joseph Dyer
Antonio Elias
Bob Ess
Kevin Eveker
Andrew Falcon
Kenneth Ford
Joseph R. Fragola
Louis Friedman
Robert E. Fudickar
Peter Garretson
Michael Gass
Bill Gerstenmaier
Mark Geyer
John Glenn
Mike Gold
Dan Goldin
Michael D. Griffin
Gene Grush
Jim Halsell
Jeff Hanley
Scott Horowitz
Matthew Isakowitz
Anthony Janetos
Tom Jasin
Chip Jones
Tom Jones
Tony Jones
Kent Joosten
John Karas
Mark Kinnersley
D.R. Komar
Dave Korsmeyer
Jeff Kottkamp
Donald Latham
Joo-Jin Lee
Matt Leonard
Dan Lester
Robert Lightfoot
Steve Lindsay
John M. Logsdon
Steve MacLean
Joanne Maguire
Ed Mango
John Marburger
Roland Martinez
James Maser
Steve Metschan
George E. Mueller
Elon Musk
Jack Mustard
Clive Neal
Scott Neish
Benjamin J. Neumann
Mike O'Brien
Sean O'Keefe
John Olson
Scott Pace
Anatoly Perminov
Pepper Phillips
Carle Pieters
Charles Precourt
Gary P. Pulliam
David Radzanowski
John Rather
Diane Rausch
Keith Reiley
Marcia Rieke
Joe Roche
Harrison "Jack" Schmitt
John Schumacher
John Shannon
Brewster Shaw
Milt Silvera
S. Fred Singer
George Sowers
Jim Spann
Paul Spudis
Steve Squyres
Thomas Stafford
Szymon Suckewer
Mike Suffredini
Phil Sumrall
Jeffrey P. Sutton
Mark Sykes
Keiji Tachikawa
Harley Thronson
Pat Troutman
Mark Uhran
Julie Van Kleeck
Zack Warfield
Johann-Dietrich Woerner
Tom Young
Robert Zubrin

Members of Congress

Representative Robert Aderholt
Representative John Culberson
Representative Davis
Representative Bart Gordon
Representative Parker Griffith
Representative Ralph Hall Representative
Suzanne Kosmas Representative Dennis
Kucinich
Representative Kendrick Meek
Representative Alan Mollohan
Representative Pete Olson
Representative Bill Posey
Senator John Cornyn
Senator Orrin Hatch
Senator Kay Hutchison
Senator Mel Martinez
Senator Barbara
Mikulski Senator Bill
Nelson
Senator Jeff Sessions
Senator Richard
Shelby Senator David
Vitter

Members of the Public

During the course of the Committee's inquiry and deliberations, more than 1,000 members of the public submitted comments, suggestions and questions, as well as documents for the Committee's consideration.

The Committee wishes to thank all who provided this valuable input.

APPENDIX G. COMMUNICATIONS AND PUBLIC ENGAGEMENT

The Committee undertook its task with a strong emphasis on receiving input from, and communicating openly with, the American public, the media, and a broad range of stakeholders in the spaceflight community. The Committee employed both traditional outreach activities as well an extensive array of Web-based and social media technologies in its efforts to facilitate maximum public engagement.

The Committee Chairman held seven press conferences at various locations throughout the United States. Two of these were teleconferences, enabling members of the media to dial in and participate from anywhere in the world. Transcripts and/or video of these press conferences were posted on the Committee's website.

Prior to beginning work, the Chairman also met individually with seven members of Congress—both Senators and Representatives, Republicans and Democrats, authorizers and appropriators. Members of the Committee participated in two hearings, one in the Senate and one in the House. In addition, many Members of Congress submitted written, oral, and videotaped statements to the Committee, which were subsequently posted on its website.

The Committee held seven public meetings: three in Washington, D.C.; one in Houston, TX; one in Huntsville, AL; one in Cocoa Beach, FL; and one via teleconference. Attendance ranged from 100-300 people at these events. At all but the August 12 public meeting and the October 8 public teleconference, the Committee reserved time for members of the public to make comments and ask questions. All public meetings were videotaped and aired live on NASA TV, and the Committee subsequently posted the videos to its website. All public meetings were also transcribed, with the transcripts also subsequently posted to the website. In addition to the public meetings, the committee held a series of closed preparatory meetings, fact-finding meetings, and site visits.

Source: Review of U.S. Human Spaceflight Plans Committee

Figure G-1. The homepage included several tools to enable public engagement with the Committee

The Chairman and the Executive Director/Designated Federal Official from NASA provided periodic progress reports to senior officials from NASA, the Office of Science and Technology Policy (OSTP), and the Office of Management and Budget (OMB). Weekly teleconferences were also held with staff members from NASA, OSTP and OMB to provide status reports.

The Committee's primary communications tool was its web- site. The website enabled anyone with Internet access to interact with the Committee in a variety of ways. The site provided ready access to information about the Committee and its activities, including: meeting presentations; videos and transcripts of public meetings; and background and related documents. These documents included the Committee Charter; Statement of Task; the Federal Register notices; press releases; meeting agendas; Congressional statements; and documents and comments submitted by the public.

The home page of the website also prominently featured a number of tools that enabled members of the public to contact and/or interact with the committee (see Figure G-1):

The homepage included several tools to enable public engagement with the Committee.

- "Provide a Comment or Suggestion" – This enabled members of the public to submit a 500-character comment or suggestion to the committee. Committee staff received more than 1,500 comments and/or suggestions during its activities.

- "Provide a Question and Get an Answer" – This provided a means for members of the public to submit questions to the Committee. The Committee screened questions for general appropriateness and then posted them to the website. Members of the public could then vote on questions that were posed. The Committee received over 250 questions, for which it developed answers that it then posted to the website. The Committee added a search capability to this feature to enable users to search for their questions and answers.
- "Follow the Committee's Recent Updates (Twitter)" – The Twitter "micro-blog" website provided a means for the Committee staff to send short, informal messages to members of the public who signed up to receive updates from the Committee. The Committee had over 2,000 "followers" through Twitter who elected to receive updates from the Committee. All public meetings were "live-tweeted," meaning that the Committee posted real-time public updates during presentations.
- "View our Photo Gallery (Flickr)" – The Committee shared pictures and images related to its work through Flickr, a photo-sharing website. The public could make comments on the photos and share images of the Committee's activities on their own Flickr accounts. Pictures of previous human spaceflight endeavors were also posted. The public viewed an average of about 500 pictures per day on the Committee's Flickr account. (See Figure G-2.)
- "Share Your Opinion on Topics" – The Committee posed three topics to stimulate public comment:

  What do you find most compelling about NASA's human spaceflight activities and why? (147 comments received)

  What role should international partners play in future U.S. spaceflight plans, and why? (98 comments received)

  To what extent should NASA rely on the private sector for human spaceflight-related products and services? (147 comments received)
- "Subscribe to the Committee's Updates via RSS" – Like many online publications, the Committee used Real Simple Syndication (RSS) feeds as another means of keeping the public informed about the Committee's activities and progress. The Committee staff posted RSS updates, each of which were approximately three to five sentences in length.
- "Join Us on Facebook" – The Committee staff developed a Facebook Fan page, which it filled with Committee information, pictures, and resources similar to the overall Committee website, as well as a Facebook Fan page public comment area, or "wall." The "wall" was used to disseminate daily information and answer general questions regarding the events, documents, and videos posted to the Committee website. The Committee had approximately 2,100 Facebook "Fans."
- "E-mail a Document" – Members of the public could email files to the Committee, a public engagement feature that had never previously been used on a NASA website. The Committee received over 200 files through this channel. When the sender indicated that a particular file could be shared with the public, the Committee posted it to its website.

- The individual “Meetings” pages allowed the public to view and share the videos of all the public meetings. Internet users could also “favorite” and comment on the videos as well.

The Committee’s goal in employing this broad spectrum of communication avenues was to set a new standard for openness and public interaction for endeavors of the type it was undertaking.

Source: Review of U.S. Human Spaceflight Plans Committee

Figure G-2. The Committee’s Flickr account enabled it to share photos of its activities with the public

APPENDIX H. BIBLIOGRAPHY

Aderholt, Robert, (2009) *Representative, video statement to the Committee*, United States House of Representatives, 29 July

Advisory Committee on the Future of the U.S. Space Program [Augustine Committee], "*Report of the Advisory Committee on the Future of the U.S. Space Program Executive Summary*", December 1990

Aerospace Safety Advisory Panel, *"Annual Report for 2008*", 2008

Altair Project Office, "*Altair Project Status*", NASA Johnson Space Center, 24 June 2009

Altair Project Office, "*Appendix A—Altair LDAC-3 Spacecraft Configuration*", NASA Johnson Space Center, 6 July 2009

Altair Project Office, "*Appendix B —Altair Configuration and Design Development"*, NASA Johnson Space Center, 6 July 2009

Altair Project Office, Appendix C—Altair Operations Concept and Design Reference Missions, Requirements Analysis Cycle 1 Baseline, NASA Johnson Space Center, 31 March 2009; *International Traffic in Arms Regulations (ITAR) controlled*

Altair Project Office, "*Appendix D—Altair Operations Timeline Document*", NASA Johnson Space Center, 6 July 2009

Altair Project Office, *Ares V Mission Planner's Guide (Draft)*, NASA Marshall Space Flight Center, November 2008

Ares I Crew Launch Vehicle System Description Preliminary Design Review (PDR) Submittal, NASA, Sensitive But Unclassified

Ares I Integrated Vehicle Design Definition Document CxP 72070 Rev B, 2 March 2009, *Sensitive But Unclassified and International Traffic in Arms Regulations (ITAR) controlled*

Ares V Integrated Vehicle Design Definition Document, Phase A, Cycle 1 Configuration, CxP 72359, NASA, 27 April 2009; Sensitive But Unclassified

Ares V Operational Concepts Document Draft, CxP 72336 and Errata Sheet, NASA Marshall Space Flight Center, 27 April 2009

Ares V Phase A Cycle 1 System Description Document (SDD), Baseline Draft, NASA, 27 April 2009; *Sensitive But Unclassified and International Traffic in Arms Regulations (ITAR) controlled*

Astronaut Office, "*Radiation and Astronaut Health Risk*", NASA Johnson Space Center

"*ATK Presentation to the Review of U.S. Human Space Flight Plans Committee"*, ATK Space Systems, 21 July 2009

Baccus, Ronald, NASA Johnson Space Center, Corliss, Jim, NASA Langley Research Center, Hixson, Robert, Lockheed Martin, "*Landing Assessments, Establishing the DAC-3 Landing Architecture"*

Bearden, David and Hart, Matthew, NASA Advanced Programs Directorate, Skratt, John, Space Launch Projects, *Human Rated Delta IV Heavy Study Constellation Architecture Impacts*, NASA and The Aerospace Corporation, 1 June 2009

Cabana, Robert, (2009) NASA Kennedy Space Center Director, "*Human Spaceflight Review Kennedy Space Center*", NASA, 30 July

Cantwell, Elizabeth, (2009) Deputy Associate Laboratory Director, National Security, *"Life and Physical Sciences in Microgravity and Partial Gravity",* Oak Ridge National Laboratory, 5 August

Chilton, Jim, Vice President and Program Manger, Exploration Launch Systems, *"Boeing Human Spaceflight Capability Overview"*, Boeing, 24 June 2009

Cline, Lynn, (2009) Deputy Associate Administrator of Space Operations Mission Directorate, *"NASA Expendable Launch Services Current Use of EELV"*, NASA Headquarters, 17 June

Cline, Lynn, (2009) *"Space Operations Mission Directorate Program Review"*, NASA Headquarters, 20 May

Coats, Mike, NASA Johnson Space Center Director, *"Johnson Space Center Perspectives for Augustine Review Panel"* [White Paper], NASA Johnson Space Center, July 2009

Coats, Mike, (2009) *"Johnson Space Center Perspectives, Augustine Review Panel",* NASA Johnson Space Center, 28 July

Collection of 10 papers on Constellation Program Near Earth Object (NEO) Mission Study Collection of 33 papers on in-space propellant transfer

Collection of 64 published papers on Other Microgravity Research Facilities

Collection of 72 International Space Station published papers

Columbia Accident Investigation Board (CAIB), *CAIB Report* Volume 1, August 2003

Commercial Spaceflight Federation, *Commercial Spaceflight in Low Earth Orbit is the Key to Affordable and Sustainable Exploration Beyond . . .*, 29 June 2009

Committee on the Evaluation of Radiation Shielding for Space Exploration, National Research Council, *Managing Space Radiation Risk in the New Era of Space Exploration*, The National Academies, 2008

Committee on the Rationale and Goals of the U.S. Civil Space Program, National Research Council, *America's Future in Space: Aligning the Civil Space Program with National Needs,* The National Academies, 2009

Committee to Review NASA's Exploration Technology Development Program, Aeronautics and Space Engineering Board, National Research Council, *A Constrained Space Exploration Technology Program: A Review of NASA's Exploration Technology Development Program*, The National Academies, 2008

Committee on the Scientific Context for Exploration of the Moon, National Research Council, *Scientific Context for the Exploration of the Moon Final Report*, The National Academies, 2007

Congressional Budget Office, *"The Budgetary Implications of NASA's Current Plan for Space Exploration"*, April 2009

Conley, Cassie, (2009) Planetary Protection Officer, *"Planetary Protection and Human Exploration",* NASA, 9 July

"Constellation Alternate Scenario Early Ares V Dual Launch Options", NASA, 28 July 2009

"Constellation Launch Vehicles Overview Part 1", NASA, 29 July 2009

Constellation Program System Requirements for the Orion System, CxP 72000, Rev D, Errata Revision and History Page, NASA, 17 June 2009

Constellation Program System Requirements for the Orion System, CxP 72000, Rev D—Pre DQA, NASA, 17 June 2009

Cooke, Doug, (2009) Associate Administrator of Exploration Systems Mission Directorate, *"Commercial Crew and Cargo Program Overview"*, NASA, 17 June

Cooke, Doug, (2009) "*Exploration Systems Mission Directorate: FY 2010 Budget Request to Congress*", NASA Headquarters, 14 May

Cooke, Doug, (2009) "*Global Exploration Strategy (GES): Highlights on Progress and Future Opportunities*", NASA, 16 June

Cooke, Doug and Hanley, Jeff, (2009) "*Review of Human Spaceflight Plans Architectural Drivers*", NASA, 16 June

Cooke, Doug and Hanley, Jeff, (2009) "*Review of Human Spaceflight Plans Constellation Overview*", NASA, 17 June

Cooke, Doug and Hanley, Jeff, (2009) "*Review of Human Spaceflight Plans Implementation*", NASA, 17 June

Coonce, Tom and Gunderson, Johanna, (2009) Office of Program Analysis and Evaluation, "*Cost and Schedule Growth and Containment at NASA*", NASA Headquarters, June

Cornyn, John, (2009) Senator, video statement to the Committee, United States Senate, 28 July

Covey, Dick, (2009) President and CEO, United Space Alliance, "United Space Alliance (USA), *Response to the Review of the U.S. Human Space Flight Plans Committee",* USA, 21 July

Craig, Douglas, *"ESR&T Support of the Assessment of Alternatives (AOA) - Development of a Modular Reusable Advanced Technology Architecture"*, NASA, March 2005; For NASA Internal Use Only—PreDecisional

Crew Exploration Vehicle Project Flight Dynamics Team, *Trajectory Design and Performance Requirements for Human Lunar Missions Using the Mission Assessment Post-Processor (MAPP) Final Report*, NASA Johnson Space Center, February 2009; Export controlled information

Cross, Robert and Turner, John, (2009) "*Constellation Program LOC/LOM [Loss of Crew/Loss of Mission] Process and Status*", NASA, 6 April

Cucinotta, Francis, "*Radiation Protection Assessment for One-Year ISS Missions*", NASA

Cucinotta, Francis, Clowdsley, Martha, Kim, Myunghee, Qualls, Garry, Simonsen, Lisa, Sulzman, Frank, Zapp, Neal, (2008) "*Phase III Constellation Architecture Trade Study Space Radiation Analyses for Lunar Outpost Variations in Risk Postures to Enable Exploration Final Data Package*", NASA, 12-14 January

Cucinotta, Francis, Kim, Myung-Hee, Willingham, Veronica, George, Kerry Physical and Biological (2008) *Organ Dosimetry Analysis for International Space Station Astronauts,* Radiation Research Society,

Culberson, John, (2009) video statement to the Committee, United States House of Representatives, 28 July

Culbertson, Frank (2009) "*Oribtal's ISS Commercial Resupply Service*", Orbital, 17 June

Defense Contract Management Agency Industrial Analysis Center, *Liquid Rocket Engines Industrial Capability Assessment*, Department of Defense, 25 September 2006, Business Sensitive, Government Proprietary— For Official Use Only

"*Depot Architecture for Lunar Exploration and Beyond*", NASA, 9 July 2009

Dordain, Jean-Jacques to the Augustine Committee, "*Statement by Jean-Jacques Dordain, Director General of the European Space Agency*", ESA, 17 June 2009

Drake, Bret, (2008) *Mars Design Reference Architecture 5.0 Study Executive Summary*, NASA, 4 December

Dyer, Joseph, (2009) NASA Aerospace Safety Advisory Panel (ASAP) Chair, "*ASAP Puts and Takes*", NASA, 14 July

Ess, Bob, (2009) "*Ares I-X Status*", NASA, 30 July

Evans, Cynthia and Robinson, Julie, (2008) Office of the International Space Station Program Scientist, NASA Johnson Space Center; Tate-Brown, Judy, Thuman, Tracy and Crespo-Richey, Jessica, Engineering and Science Contract Group; Bauman, David and Rhatigan, Jennifer, NASA Johnson Space Center, International Space Station Science Research Accomplishments During the Assembly Years: An Analysis of Results from 2000-2008 Executive Summary, Houston, Texas, December

Eveker, Kevin, (2009) "*The Budgetary Implications of NASA's Current Plans for Space Exploration*", Congressional Budget Office, 16 June

Exploration Systems Architecture Study (ESAS) Team, "*ESAS Executive Summary*", NASA, November 2005

Exploration Systems Mission Directorate, *Lunar Architecture Focused Trade Study Final Report, Rev A*, ESMD-RQ-0005, NASA Headquarters, 4 February 2005

Exploration Systems Mission Directorate, Lunar Architecture Broad Trade Study Final Report, Rev A, ESMD-RQ-0006, NASA Headquarters, 4 February 2005

Exploration Systems Mission Directorate, "*Lunar Vehicle/ Lunar Architecture Analysis of Alternatives (AoA)*", NASA, 8 February 2005, Pre-Decisional, NASA Internal Use Only

Exploration Transportation System Strategic Roadmap Committee, *Final Report,* NASA, May 2005

Fragola, Joseph, (2009) "The Path to a Safer Crew Launch Vehicle, The Ares I Story", Valador, 29 July

Fragola, Joseph, (2009) "*White Paper: Ares I, the Crew Launcher with Crew Safety Designed In*", Valador, Inc., June

Friedman, Louis, (2008) Roadmap Team, "*Beyond the Moon, A New Roadmap for Human Space Exploration in the 21st Century*", The Planetary Society, November

Gass, Michael, (2009) President and CEO of United Launch Alliance (ULA), "*Briefing to the Review of U.S. Human Space Flight Plans Committee*", ULA, 16 June, ULA Proprietary Information

Gass, Michael, (2009) President and CEO of United Launch Alliance (ULA), "*Briefing to the Review of U.S. Human Space Flight Plans Committee*", ULA, 17 June

Gerstenmaier, Bill, (2009) Associate Administrator for Space Operations Mission Directorate, "*International Space Station Extension and Utilization Planning*", NASA Headquarters, 9 June

Geyer, Mark, (2009) Orion Project Office Manger, "*Review of Human Spaceflight Plans Orion Crew Exploration Vehicle*", NASA, 28 July

"*Global Exploration Strategy (GES) Framework: Executive Summary*", NASA and Other Space Agencies, May 2007

Gordon, Bart, (2009) Representative, letter to Mr. Norm Augustine, United States House of Representatives, 17 June

Griffith, Parker, (2009) Representative, video statement to the Committee, United States House of Representatives, 29 July

Griffin, Michael, (2005) NASA Administrator, and Sega, Ronald, Department of Defense Executive Agent for Space, to John Marburger III, Director, Office of Science and Technology Policy, Washington, DC, 5 August

Hall, Ralph, (2009) Representative, "*Statement of the Honorable Ralph Hall (R-TX), Ranking Member, U.S. House Committee on Science and Technology*", United States House of Representatives, 17 June

Hanley, Jeff, (2009) Constellation Program Manger, "*NASA Constellation Projects Closing Comments*", NASA, 30 July

Hanley, Jeff, (2009) "*NASA Constellation Projects Introduction*", NASA, 28 July

Hawes, W. Michael, (2009) "*Summary Description of Previous Studies*", NASA, 17 June

"*Human Exploration of Mars, Design Reference Architecture 5.0*", NASA, 29 July 2009

"*Human Research Program Utilization Plan for the International Space Station*", NASA

"*Human Space Flight Capabilities, pursuant to Section 611(a) of the NASA Authorization Act of 2008 (P.L. 110-422)*", NASA, April 2009

Hutchison, Kay Bailey, Senator, (2009) "*Statement by Senator Kay Bailey Hutchison for Submission to Augustine Review Panel*", United States Senate, 17 June

Hutchison, Kay Bailey, (2009) Senator, "*Statement by Senator Kay Bailey Hutchison for Submission to Augustine Review Panel Tuesday, July 28, 2009*", United States Senate, 28 July

Impacts of Shuttle Extension, pursuant to Section 611(e) of the NASA Authorization Act of 2008 (P.L. 110-422), NASA, April 2009

Innovative Partnerships Program, Spinoff: 50 Years of NASA-Derived Technologies (195 8-2008), NASA, 2008

"*International Space Station, Research Facilities on ISS, International Partner Plans for Research*", NASA, June 2009

Janetos, Anthony, (2009) "*Earth Science and Applications from Space, National Imperatives for the Next Decade*", 5 August

Jones, Tom, (2009) "*The Asteroid Opportunity: Human Exploration of Near Earth Objects*", Association of Space Explorers and NASA Advisory Council, 9 July

Joosten, Kent, "*Constellation Lunar Architecture Overview*", NASA, 9 July 2009

Kinnersley, Mark, (2009) Director of Business Development, Orbital System and Exploration Division, "*EADS Astrium Statement to the Review of U.S. Human Spaceflight Plans Committee*", EADS Astrium, 5 August

Kinnersley, Mark, (2009) "*Statement by EADS Astrium*", EADS Astrium, 5 August

Kosmas, Suzanne, Representative, "*Statement of Congresswoman Suzanne Kosmas (FL-24), The Review of U.S. Human Space Flight Plans Committee*", United States House of Representatives, 30 July 2009

Kottkamp, Jeff, (2009) Lieutenant Governor of Florida, statement in person to the Committee, 30 July

Kucinich, Dennis, (2009) Representative, letter to Mr. Norm Augustine, United States House of Representatives, 4 August

Le Gall, Jean-Yves, (2009) Chairman and CEO of Arianespace, "*Presentation to the United States Human Space Flight Plans Review Committee*", Arianespace, 5 August

Leonard, Matt, (2009) Lunar Surface Systems, Deputy Project Manager, "*Lunar Surface Systems (LSS) Briefing to Augustine Committee*", 9 July

Lester, Dan, (2009) "*Future of Human Space Flight and Astronomy*", the University of Texas, 9 July

Lightfoot, Robert, (2009) NASA Marshall Space Flight Center (MSFC) Acting Center Director, "*Marshall Space Flight Center Launching the Future of Exploration and Science*", NASA MSFC, 29 July

Lindsey, Steve, (2009) Chief of the Astronaut Office, "*Human Spaceflight Review Astronaut Office Perspective*", NASA Johnson Space Center, 28 July

MacLean, Steve, (2009) President, Canadian Space Agency (CSA), "*CSA-NASA Human Space Flight Cooperation*", CSA, 16 June

Manella, Rick (2009) "*Shuttle Derived Heavy Lift Launch Vehicle*", NASA, 25 June

Maguire, Joanne, (2009) Executive Vice President, "*Review of the U.S. Human Space Flight Plans Committee, Committee Discussion*", Lockheed Martin Space Systems Company, 22 July

Mango, Ed, (2009) *"Constellation Space Transportation Planning Office"*, NASA, 30 July

Marburger, John, (2009) former Science Advisor to President George W. Bush and Director, Office of Science and Technology Policy (2001-2009), *"Review of U.S. Human Space Flight Plans Committee Remarks on the Background for the Vision for Space Exploration"*, 5 August

Mars Architecture Steering Group, *Human Exploration of Mars Design Reference Architecture 5.0*, NASA/SP2009-566, NASA, July 2009

Mars Architecture Steering Group, *Human Exploration of Mars Design Reference Architecture 5.0 Addendum*, NASA/SP-2009-566-ADD, NASA, July 2009

Martinez, Mel, Senator, (2009) *"Testimony of U.S. Senator Mel Martinez"* [and accompanying video], United States Senate, 30 July

Maser, James, President, Pratt & Whitney Rocketdyne, (2009) *"Briefing to the Review of U.S. Human Space Flight Plans Committee*", 8 July

Meek, Kendrick, (2009) Representative, *"Statement of U.S. Representative Kendrick Meek (FL- 17)"*, United States House of Representatives, 30 July

Metschan, Steve, (2009) *"DIRECT—Safer, Simpler and Sooner"*, DIRECT, 17 June

Multilateral Coordination Board, *International Space Station Lessons Learned as Applied to Exploration*, NASA, Canadian Space Agency (CSA), European Space Agency (ESA), Japanese Aerospace Exploration Agency (JAXA), Roscosmos, 22 July 2009

Musk, Elon, (2009) *"COTS Status Update & Crew Capabilities"*, SpaceX, 16 June

Musk, Elon, (2009) "COTS Status Update & Crew Capabilities", SpaceX, 17 June

Mustard, Jack, (2009) Mars Exploration Program Analysis Group Chair, *"Human and Robotic Science on Mars"*, NASA, 9 July

National Academy of Sciences, *"The Scientific Context for the Exploration of the Moon Executive Summary"*, Washington, DC, 2007

NASA Authorization Act of 2005 (Senate Resolution 1281), 109th Congress, December 2005

NASA Authorization Act of 2008 (House Resolution 6063), 110th Congress, October 2008

NASA Office of External Relations, *Background on NASA International Cooperation in Human Space Flight Activities*, NASA Headquarters, June 2009

NASA Office of External Relations, *"Non-NASA Government Activities and Needs Bearing on Human Space Flight Planning"*, NASA Headquarters, June 2009

NASA Organizational Model Evaluation Team, Process, Analysis and Recommendations, NASA March 2005

NASA Procedural Requirements 8705.2B, *Subject: Human-Rating Requirements for Space Systems*, May 2008-May 2013

NASA Procedural Requirements 8715.3, Subject: *General Safety Program Requirements*, April 2009 - April 2014

"NASA Report to Congress Regarding a Plan for the International Space Station National Laboratory", NASA, May 2007

"NASA Research and Utilization Plan for the International Space Station Executive Summary", NASA

NASA Space Shuttle Workforce Transition Strategy pursuant to FY 2008 Consolidated Appropriations Act (P.L. 110-161), July 2009 Update

NASA Strategic Roadmap Summary Report, NASA, 22 May 2005

"NASA Technology Development, Supporting the Mission and the Planet", NASA, 31 July 2009

"NASA Workforce Transition Strategy, Space Shuttle and Constellation Workforce Focus, Biannual Report— Second Edition", NASA, October 2008

National Council on Radiation Protection and Measurements, *Information Needed to Make Radiation Protection Recommendations for Space Missions Beyond Low-Earth Orbit NCRP* Report No. 153, 15 November 2006

Nature Review Cancer, *"Heavy Ion Carcinogenesis and Human Space Exploration"*, Advance Online Publication, 2 May 2008

Neal, Clive, (2009) Lunar Exploration Analysis Group Chair, *"Lunar Studies w/Astronauts and/or Robots"*, University of Notre Dame, 9 July

Nelson, Bill, (2009) Senator, video statement to the U.S. Human Space Flight Plans Committee, United States Senate, 30 July

O'Brien, Michael, (2009) Assistant Administrator for External Affairs, *"Overview of International Partnerships"*, NASA, 16 June

Olson, John, (2009) *"Dual Ares V Lunar Architecture Option: A Quick Look at a Blue Sky Derived Concept"*, NASA, 9 July

Office of Biological and Physical Research, *"Requirements for Human Subjects in Exploration Research Workshop"*, NASA Headquarters, 12-13 May 2004

Office of Safety and Mission Assurance, Human-Rating Requirements for Space Systems, *NASA Procedural Requirements (NPR)* 8705.2B, NASA, 6 May 2008

Office of Under Secretary of Defense, Acquisition, Technology and Logistics Industrial Policy, Draft Solid Rocket Motor (SRM) Industrial Capabilities Report to Congress, Department of Defense, February 2009, Proprietary Business Sensitive Information—Not Releasable Outside the Government

Olson, Pete, Representative, (2009) *"The Honorable Pete Olson (R - TX) Ranking Member, U.S. House Space and Aeronautics Subcommittee, Remarks to the Augustine Panel"*, United States House of Representatives, 17 June

Olson, Pete, (2009) Representative, video statement to the Committee, United States House of Representatives, 28 July

OSP [Orbital Space Plane] —*ELV [Expendable Launch Vehicle] Human Flight Safety Certification Study Report, Volume 1*, NASA, March 2004, International Traffic in Arms Regulations (ITAR) and Export controlled

OSP [Orbital Space Plane] —*ELV [Expendable Launch Vehicle] Human Flight Safety Certification Study Report Data Appendices,* NASA, March 2004, Proprietary, International Traffic in Arms Regulations (ITAR) and Export controlled

Patton, Jeff, *"Program P401 Launch Vehicle Overview TIM [Technical Interchange Meeting] #1"*, United Launch Alliance (ULA), 5 May 2008, Bigelow Aerospace Proprietary Information and ULA Proprietary Information

Perminov, Anatolii, Head of the Russian Federation Space Agency, *"ISS Program International Cooperation"*, Russian Federation Space Agency (Roscosmos), 17 June 2009

Phillips, Pepper, (2009) "*Ground Operations*", NASA, 30 July

Posey, Bill, Representative, (2009) "*Statement of U.S. Representative Bill Posey (FL-15)*", United States House of Representatives, 30 July

President's Commission on Implementation of United States Space Exploration Policy [Aldridge Report], "*A Journey to Inspire, Innovate and Discover Executive Summary*", Executive Office of the President, June 2004

"Project 401 Technical Interchange Meeting #1", Lockheed Martin, 5 May 2008, Bigelow *Aerospace Proprietary Information and Lockheed Martin Proprietary/Export Controlled Data*

"Project 401 Technical Interchange Meeting #2", Lockheed Martin, 5 May 2008, Bigelow Aerospace Proprietary Information and Lockheed Martin Proprietary/Export Controlled Data

Pulliam, Gary, (2009) Vice President of Civil and Commercial Operations, The Aerospace Corporation, "*Initial Summary of Human Rated Delta IV Heavy Study*", The Aerospace Corporation, 17 June

Radzanowski, David, (2009) Deputy Associate Administrator, Program Integration, "*Background on the Budget, Review of Human Space Flight Plans*", NASA Headquarters, 9 June

Rieke, Marcia, (2009) Professor of Astronomy, "*Opportunities and Challenges in Astronomy and Astrophysics*", University of Arizona, 5 August

Rutkowski, Brian, (2009) Office of Program Analysis and Evaluation, "*Schedules: Predicting the Next KDP Milestone* (Plans vs. Actuals), June 2009", NASA Headquarters, June

Sessions, Jeff, Senator, (2009) letter to the U.S. Human Space Flight Plans Committee, United States Senate, 21 July

Shannon, John, (2009) Manager, Space Shuttle Program, "*Shuttle- Derived Heavy Lift Launch Vehicle*", NASA, 17 June

Shannon, John, "*Space Shuttle Program Overview*", NASA, 16 June 2009

Shelby, Richard, Senator, (2009) letter to the U.S. Human Space Flight Plans Committee, United States Senate, 29 July

Sowers, George, Vice President, Business Development and Advanced Programs, United Launch Alliance (ULA), "*Atlas and Delta Human Rating Overview*", 24 June 2009, ULA Proprietary Information

Space and Life Sciences Directorate, Net Habitable Volume Verification Method, JSC 63557, NASA Johnson Space Center, 9 April 2009

Space and Life Sciences Directorate, "*Space Radiation and Exploration—Information for the Augustine Committee Review*", NASA Johnson Space Center, 14 July 2009 Space and Life Sciences Directorate, "Space Radiation and Exploration—Information for the Augustine Committee Review, Supplementary Material on Lunar Regolith Shielding", NASA Johnson Space Center, 14 July 2009

Spann, James, (2009) "*Science from the Moon*", NASA Marshall Space Flight Center, 9 July

Sponable, Jesse, (2009) "*Reusable Space Systems, 21st Century Challenges*", Defense Advanced Research Projects Agency, 17 June

Strategy and Architecture Office, ESA Integrated Exploration Architecture Study, European Space Agency (ESA)

Squyres, Steven, (2009) "*Briefing to the NASA Review of U.S. Human Space Flight Plans Committee*", 5 August

Suffredini, Michael, (2009) International Space Station Program Manager, "*ISS Commercial Resupply Services*", NASA, 17 June

Suffredini, Michael, (2009) "*International Space Station Presentation to Augustine Commission*", NASA, 22 July

Sykes, Mark, (2009) Director, Planetary Science Institute and Chair, NASA Small Bodies Assessment Group, "*NEO Science and Human Space Activity*", Planetary Science Institute and NASA, 9 July

Synthesis Group, "*America at the Threshold: America's Space Exploration Initiative Executive Summary*", Executive Office of the Vice-President, May 1991

Tachikawa, Keiji, (2009) President, Japanese Aerospace Exploration Agency (JAXA), "Statement of Dr. Keiji Tachikawa, President, Japanese Aerospace Exploration Agency (JAXA) before the Review of U.S. Human Space Flight Plans Committee", JAXA, 16 June

"Technical Infusion for Exploration", NASA, 7 August 2009

"*Technology Development for Cryogenic Propellant Depots, Information for the Review of U.S. Human Space Flight Plans Committee"*, NASA, 4 August 2009

Thronson, Harley, Associate Director for Advanced Concepts and Planning, Astrophysics Science Division, *"On-Orbit Satellite Upgrade and Maintenance: Astronauts and Robots"*, NASA Goddard Space Flight Center, 9 July 2009

"U.S. National Space Policy", Executive Office of the President, 31 August 2006

"U.S. Space Transportation Policy Fact Sheet", Executive Office of the President, 6 January 2005

The Vision for Space Exploration [also included: "Background", *"Bringing the Vision to Reality", "Goals and Objectives*", the speech and "White House Fact Sheet: A Renewed Spirit of Discovery"], Executive Office of the President, February 2004

Vitter, David, Senator, (2009) *"Statement of Senator David Vitter"*, United States Senate, 17 June

Yoder, Geoffrey, *"COTS Updated Status",* NASA, 29 July 2009, Sensitive But Unclassified—For NASA Internal Use Only

Zapata, Edgar, Operations Analysis, *"NASA's Historical Moment- Goals, Budgets, Improvement & Safety"*, NASA Kennedy Space Center, 11 May 2009

Zapata, Edgar, Operations Analysis, *"NASA Re-booted"*, NASA Kennedy Space Center, 12 May 2009

Zapata, Edgar, (2009) *Operations Analysis, "NASA Strategic Scenario Analysis"*, NASA Kennedy Space Center, 10 June

Zubrin, Robert, (2009) *"Accepting the Challenge Before Us",* Pioneer Astronautics, 5 August

Zubrin, Robert, (2009) *"Mars Direct: Humans to the Red Planet within a Decade"*, Pioneer Astronautics, 5 August

GLOSSARY ACRONYMS AND ABBREVIATIONS

CAIB	Columbia Accident Investigation Board
COTS	Commercial Orbital Transportation Services
CSA	Canadian Space Agency
DDT&E	Design, Development, Test, & Evaluation
DOD (or DoD)	Department of Defense
EDL	Entry, Descent and Landing
EDS	Earth Departure Stage
E ELV	Evolved Expendable Launch Vehicle
ESA	European Space Agency
ESAS	Exploration Systems Architecture Study
ESMD	Exploration Systems Mission Directorate
EVA	extra-vehicular activity (spacewalk)
FAA	Federal Aviation Administration
FY	Fiscal Year
GDP	Gross Domestic Product
G ES	Global Exploration Strategy
INKSNA	Iran, North Korea, and Syria Nonproliferation Act
ISS	International Space Station
ISRO	Indian Space Research Organisation
ITAR	International Traffic in Arms Regulations
JAXA	Japan Aerospace Exploration Agency
KSC	Kennedy Space Center
KM	Kilometers
LEO	low-Earth orbit
LH2	liquid hydrogen
LOX	liquid oxygen
mt	metric ton
NASA	National Aeronautics and Space Administration
NEO	near-Earth object
NSS	national security space
NTR	nuclear thermal rocket
OMB	Office of Management and Budget
OSTP	Office of Science & Technology Policy
PDR	Preliminary Design Review
PRA	probabilistic risk assessment
PRC	People's Republic of China
SDR	System Design Review
SEI	Space Exploration Initiative
SRB	Solid Rocket Booster
SRM	Solid Rocket Motor
SSME	Space Shuttle Main Engine
STEM	Science, Technology, Engineering and Mathematics

STS	Space Transportation System (Shuttle)
TLI	Trans Lunar Injection
TPS	Thermal Protection System
TRL	Technology Readiness Level
ULA	United Launch Alliance
VSE	Vision for Space Exploration

End Notes

[1] Norman R. Augustine, Augustine's Laws (Washington, D.C.: American Institute of Aeronautics and Astronautics), 1986.

In: Future of U.S. Human Spaceflight: Background and Issues ISBN: 978-1-61668-774-8
Ed: Derek A. Warren and Bridget D. Conway

Chapter 2

UNITED STATES CIVILIAN SPACE POLICY PRIORITIES: REFLECTIONS 50 YEARS AFTER SPUTNIK*

Deborah D. Stine

SUMMARY

The "space age" began on October 4, 1957, when the Soviet Union (USSR) launched Sputnik, the world's first artificial satellite. Some U.S. policymakers, concerned about the USSR's ability to launch a satellite, thought Sputnik might be an indication that the United States was trailing behind the USSR in science and technology. The Cold War also led some U.S. policymakers to perceive the Sputnik launch as a possible precursor to nuclear attack. In response to this "Sputnik moment," the U.S. government undertook several policy actions, including the establishment of the National Aeronautics and Space Administration (NASA) and the Defense Advanced Research Projects Agency (DARPA), enhancement of research funding, and reformation of science, technology, engineering and mathematics (STEM) education policy.

Following the "Sputnik moment," a set of fundamental factors gave "importance, urgency, and inevitability to the advancement of space technology," according to an Eisenhower presidential committee. These four factors include the compelling need to explore and discover; national defense; prestige and confidence in the U.S. scientific, technological, industrial, and military systems; and scientific observation and experimentation to add to our knowledge and understanding of the Earth, solar system, and universe. They are still part of current policy discussions and influence the nation's civilian space policy priorities—both in terms of what actions NASA is authorized to undertake and the appropriations each activity within NASA receives.

---

* This is an edited, reformatted and augmented version of a CRS Report for Congress publication dated February 2009.

Further, the United States faces a far different world today. No Sputnik moment, Cold War, or space race exists to help policymakers clarify the goals of the nation's civilian space program. The Hubble telescope, *Challenger* and *Columbia* space shuttle disasters, and Mars exploration rovers frame the experience of current generations, in contrast to the Sputnik launch and the U.S. Moon landings. As a result, some experts have called for new 21st century space policy objectives and priorities to replace those developed 50 years ago.

On October 15, 2008, the NASA Authorization Act of 2008 (P.L. 110-422) was signed into law. This act authorized appropriations for FY2009, and prohibited NASA from taking any steps prior to April 30, 2009, that would preclude the President and Congress from being able to continue to fly the Space Shuttle past 2010. During the 111th Congress, policymakers may discuss another authorization bill for future years, and identify priorities for civil space exploration. Little is known about the Obama Administration's space policy priorities, and the degree to which they agree with Bush Administration policies.

If policymakers identify priorities for U.S. civil space exploration, this might help Congress determine the most appropriate balance of funding for NASA's programs during its authorization and appropriation process. For example, if Congress believes that national prestige should be the highest priority, they may choose to emphasize NASA's human exploration activities, such as establishing a Moon base and landing a human on Mars. If they consider scientific knowledge the highest priority, Congress may emphasize unmanned missions and other science-related activities as NASA's major goal. If international relations are a high priority, Congress might encourage other nations to become equal partners in actions related to the International Space Station. If spinoff effects, including the creation of new jobs and markets and its catalytic effect on math and science education, are Congress' priorities, then they may focus NASA's activities on technological development and linking to the needs of business and industry, and expanding its role in science and mathematics education.

Current U.S. space policy is based on a set of fundamental factors which, according to an Eisenhower presidential committee, "give importance, urgency, and inevitability to the advancement of space technology."[1] These factors were developed fifty years ago as a direct result of the Soviet Union's (USSR) launch of the first artificial satellite, Sputnik. This launch began the "space age" and a "space race" between the United States and USSR.

The four factors are the compelling need to explore and discover; national defense; prestige and confidence in the U.S. scientific, technological, industrial, and military systems; and scientific observation and experimentation to add to our knowledge and understanding of the Earth, solar system, and universe.[2] They are still part of current policy discussions and influence the nation's civilian space policy priorities—both in terms of what actions NASA is authorized to undertake and the appropriations each activity within NASA receives.

NASA has active programs that address all four factors, but many believe that it is being asked to accomplish too much for the available resources. An understanding of how policy decisions made during the Sputnik era influence U.S. space policy today may be useful as Congress considers changing that policy. The response of Congress to the fundamental question, "Why go to space?," may influence NASA's programs, such as its earth-observing satellites, human exploration of the Moon and Mars, and robotic investigation of the solar system and wider universe as well as its policies on related activities, including spinoff technological development,[3] science and mathematics education, international relations, and commercial space transportation.

This report describes Sputnik and its influence on today's U.S. civilian space policy, the actions other nations and commercial organizations are taking in space exploration, and why the nation invests in space exploration and the public's attitude toward it. The report concludes with a discussion of possible options for future U.S. civilian space policy priorities and the implication of those priorities.

SPUTNIK AND AMERICA'S "SPUTNIK MOMENT"

On October 4, 1957, the USSR launched Sputnik, the world's first artificial satellite. Sputnik (Russian for "traveling companion") was the size of a basketball and weighed 183 pounds (see Figure 1). Sputnik's launch and orbit[4] still influences policy decisions 50 years later.

The USSR's ability to launch a satellite ahead of the United States led to a national concern that the United States was falling behind the USSR in its science and technology capabilities and thus might be vulnerable to a nuclear missile attack.[5] The resulting competition for scientific and technological superiority came to represent a competition between capitalism and communism.

Both the 85th Congress and President Eisenhower undertook an immediate set of policy actions in response to the launch of Sputnik. Congress established the Senate Special Committee on Space and Astronautics on February 6, 1958, and the House Select Committee on Science and Astronautics on March 5, 1958—the first time since 1892 that both the House and Senate took action to create standing committees on an entirely new subject. Each committee was chaired by the Majority Leader. The Preparedness Investigating Subcommittee of the Senate Armed Services Committee was also active in analyzing the nation's satellite and missile programs.[6]

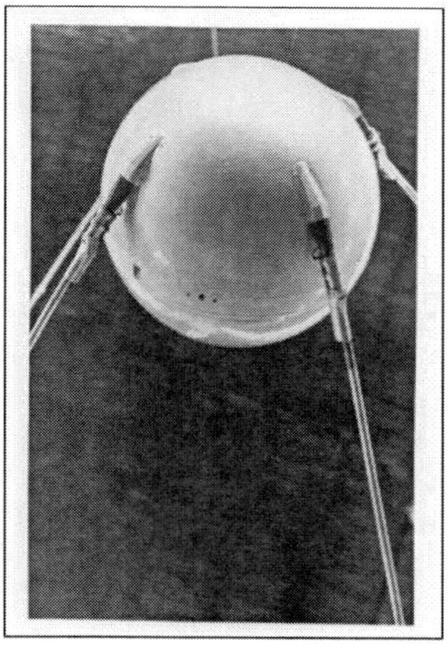

Source: NASA

Figure 1. Sputnik

Multiple congressional hearings were held in the three months following Sputnik, and President Eisenhower addressed the nation to assure the public that the United States was scientifically strong and able to compete in space. Within 10 months after Sputnik's launch, the Eisenhower Administration and Congress took actions that

- established the National Aeronautics and Space Administration (NASA) through the National Aeronautics and Space Act (P.L. 85-568),[7]
- established the Defense Advanced Research Projects Agency (DARPA) within the Department of Defense through DOD Directive 5105.15 and National Security - Military Installations and Facilities (P.L. 85-325),[8]
- increased its appropriation for the National Science Foundation to $134 million, nearly $100 million higher than the previous year,[9] and
- reformed elementary, secondary, and postsecondary science and mathematics education (including gifted education) and provided incentives for American students to pursue science, technology, engineering, and mathematics postsecondary degrees via fellowships and loans through the National Defense Education Act (P.L. 85-864).[10]

Figure 2 provides a timeline of the some of the major policy events in the year following the Sputnik launch.

When people today speak of a "Sputnik moment," they often refer to a rapid national response that quickly mobilizes major policy change as opposed to a response of inaction or incremental policy change. The term is also used to question inaction—as in whether or not the nation is prepared to respond to a challenge without an initiating Sputnik moment.

Why Was Sputnik so Influential?

The Sputnik launch captured the public's attention at a time of heightened U.S. tension regarding the threat posed by the USSR and communism. Societal focus on civil defense, including "duck and cover" drills and the establishment of some personal bomb shelters, predisposed the nation towards identifying the potential threat posed by the Sputnik launch.[11] In this climate, many Americans became concerned that if the USSR could launch a satellite into space, it could also launch a nuclear missile capable of reaching the United States.[12]

The Sputnik launch was immediately viewed as a challenge to U.S. scientific and technological prowess. The Soviet Union launched both Sputnik and Sputnik 2 before the United States was able to attempt a satellite launch.[13] Additionally, the Soviet launch was of a far heavier satellite than the U.S. had planned.[14] The net result of the Sputnik launch was called a "Pearl Harbor for American Science"—a sign that the United States was falling behind the USSR in science and technology.[15] The ensuing competition in scientific and technological skills came to represent a competition to determine the political superiority of capitalism versus communism.

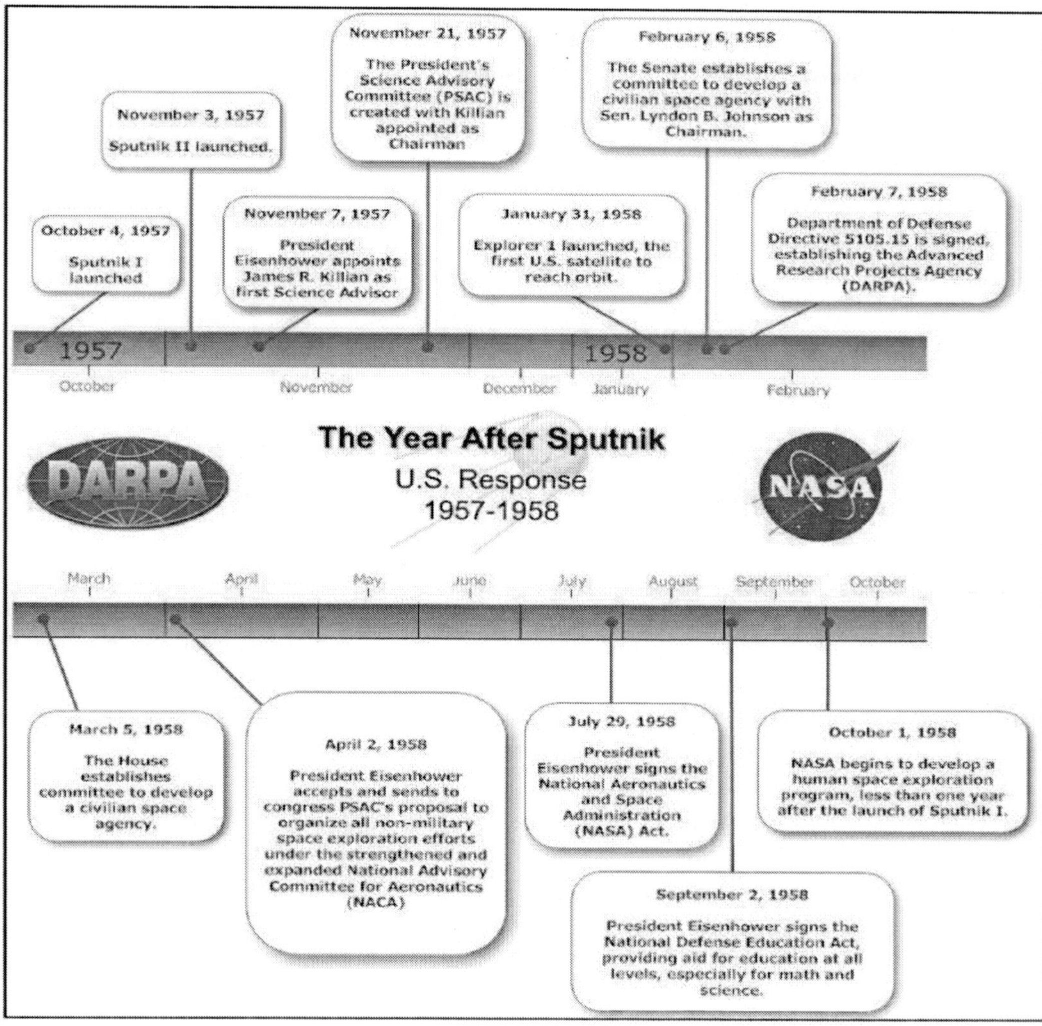

Source: Association of American Universities
Notes: DARPA was also established by Congress in P.L. 85-325.

Figure 2. Timeline of Select Policy Events in the Year Following the Sputnik Launch

The Senate Majority Leader at the time, future President Lyndon B. Johnson, illustrated the concern of many Americans in his own observations of the night sky: "Now, somehow, in some new way, the sky seemed almost alien. I also remember the profound shock of realizing that it might be possible for another nation to achieve technological superiority over this great country of ours."[16]

WHY IS SPUTNIK IMPORTANT TO TODAY'S POLICIES?

The Sputnik launch prompted rapid development of new federal policies and programs. In particular, federal investment in NASA is still influenced by the Sputnik-era principles as illustrated in the Space Act, both in terms of what actions NASA is authorized to undertake and the extent to which each activity is funded.

In 2008, NASA was reauthorized for FY2009.[17] As Congress considers future reauthorization of NASA, the status of the nation's space policy, and the relative importance of the various objectives underlying this policy may become topics of debate.

The United States faces a far different world today than 50 years ago. No Sputnik moment, Cold War, or space race exists to help policymakers clarify the goals of the nation's civilian space program. The Hubble telescope, *Challenger* and *Columbia* space shuttle disasters, and Mars exploration rovers frame the experience of current generations, in contrast to the Sputnik launch and the U.S. Moon landings that form the experience of older generations.

What Are the Activities of Other Nations and the Commercial Sector in Space Exploration?

According to an analysis conducted by the Space Foundation, the global space industry in 2007 generated $251.16 billion in revenues.[18] (See Figure 3.) The United States faces a possible new set of competitors or collaborators in civilian space exploration. China, India, Japan, Russia, and Europe are taking an active role in space exploration as are commercial companies.[19] If China is the first to land humans on the Moon and establish a Moon base in the 21st century or the European Space Agency is the first to land humans on Mars, will policymakers and the public view these activities as a loss in United States status and leadership? If so, what are the policy implications? Would such activities become this century's "Sputnik moment" that would spur further investment in U.S. space exploration activities? If not, how might this affect U.S. space policy priorities?

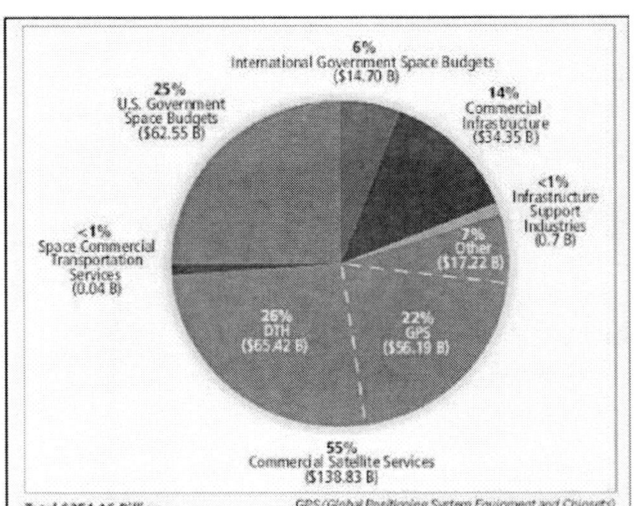

Source: Space Foundation, The Space Report: Guide to Global Space Activities, 2008

Figure 3. Global Space Activity, 2007

Future spacecraft are being developed. For example, the X-Prize Foundation Google Lunar X Prize ($30 million) invites private teams from around the world to build a robotic rover capable of landing on the Moon.[20] Virgin Galactic, currently based in California with a spaceport under construction in New Mexico, has plans for SpaceShipTwo, a six-passenger spaceliner.[21] In Europe, EADS-Astrium is developing a four-person spacecraft to make suborbital trips.[22] According to press reports, a number of venture capitalists are also planning to build spaceships or develop private space programs.[23] Should these efforts prove successful, what implications might this have for U.S. space policy priorities?

What Is the Nation's Current Civilian Space Policy?

Little is known about the Obama Administration's space policy. The Obama Administration has said that it "will restore American leadership on space issues,"[24] but only provides details regarding its military space policy. The Government Accountability Office (GAO) indicates that a decision as to whether or not to retire the space shuttle among the top 13 issues facing President Obama. The degree to which the Obama Administration agrees with the Bush Administration space policy is unknown.

During the Bush Administration, a U.S. National Space Policy defined the key objectives of defense and civilian space policy.[25] This policy incorporated key elements of the Vision for Space Exploration ("Vision"), often referred to as the Moon/Mars program. In the Vision,[26] the President directed NASA to focus its efforts on returning humans to the Moon by 2020 and eventually sending them to Mars and "worlds beyond."[27] The President further directed NASA to fulfill commitments made to the 13 countries that are its partners in the International Space Station (ISS). In the 2005 NASA authorization act (P.L. 109-155), Congress directed NASA to establish a program to accomplish the goals outlined in the Vision, which are that the United States

- Implement a sustained and affordable human and robotic program to explore the solar system and beyond;
- Extend human presence across the solar system, starting with a human return to the Moon by the year 2020, in preparation for human exploration of Mars and other destinations;
- Develop the innovative technologies, knowledge, and infrastructures both to explore and to support decisions about the destinations for human exploration; and
- Promote international and commercial participation in exploration to further U.S. scientific, security, and economic interests.[28]

More specifically, the Vision included plans, via a strategy based on "long-term affordability," to

- return the Space Shuttle safely to flight (which has been accomplished),
- complete the International Space Station (ISS) by 2010 but discontinue its use by 2017,
- phase out the Space Shuttle when the ISS is complete by 2010,

- send a robotic orbiter and lander to the Moon,
- send a human expedition to the Moon (sometime between 2015-2020),
- send a robotic mission to Mars in preparation for a future human expedition, and
- conduct robotic exploration across the solar system.[29]

NASA is developing a new spacecraft called Orion (formerly the Crew Exploration Vehicle) and a new launch vehicle for it called Ares I (formerly the Crew Launch Vehicle). An Earth-orbit capability is planned by 2014 (although NASA now considers early 2015 more likely) with the ability to take astronauts to and from the Moon following no later than 2020.

The Vision had broad implications for NASA, especially since almost all the funds to implement the initiative are expected to come from other NASA activities. Among the issues Congress is debating are the balance between NASA's exploration activities and its other programs, such as science and aeronautics research; the impact of the Vision on NASA's workforce needs; whether the space shuttle program might be ended in 2010; and if the United States might discontinue using the International Space Station.[30]

NASA stated that its strategy is to "go as we can afford to pay," with the pace of the program set, in part, by the available funding.[31] Affording such a program is challenging, however, with a 2006 National Research Council report finding "NASA is being asked to accomplish too much with too little." The report recommended that "both the executive and the legislative branches of the federal government need to seriously examine the mismatch between the tasks assigned to NASA and the resources that the agency has been provided to accomplish them and should identify actions that will make the agency's portfolio of responsibilities sustainable."[32]

WHY INVEST IN SPACE EXPLORATION?

The Table A-1 compares The National Aeronautics and Space Act of 1958 as amended ("Space Act"),[33] the oldest and most recent Presidential commission reports (Killian[34] and Aldridge[35]), the U.S. National Space Policy[36] ("Space Policy"), and the National Aeronautics and Space Administration Authorization Act of 2008 (P.L. 110-422). The analyses identify the following reasons why the United States might explore space:

- knowledge and understanding,
- discovery,
- economic growth—job creation and new markets,
- national prestige, and
- defense.

Some also include the following reasons:

- international relations, and
- education and workforce development.

Although there is broad agreement on the reasons for space exploration, there is a great deal of variation in the details. Among the chief differences in these documents are the degree to which

- discovery is the major reason for space exploration as opposed to meeting needs here on Earth;
- creation of jobs and new markets should be a major focus of NASA activities as opposed to a side effect;
- science and mathematics education and workforce development should be a goal of NASA in addition to other federal agencies; and
- relationships with other countries should be competitive or cooperative regarding space exploration.

Comparing the Aldridge Commission themes, the Space Policy goals, and the Space Act objectives on the issue of the relationship of the space program to economic growth provides some insights. While the Aldridge committee has a much broader view of the industries related to space exploration, focusing on the potential role of space exploration in job generation and new market development, the Space Act and Space Policy focus on only one sector, the aeronautical and space vehicle industry.

The two Presidential commissions have two key differences. One is the first theme outlined in the Sputnik-era Killian Committee report: "the compelling urge of man to explore and discover." This is quite different from the recent Aldridge Commission report, which, although indicating exploration and discovery should be among NASA goals, states that "exploration and discovery will perhaps *not* be sufficient drivers to sustain what will be a long, and at times risky, journey." The implication is that, today, solely responding to the challenge of going to the Moon or Mars is not sufficient to energize public support for space exploration.

The second key difference is the focus of the Aldridge Commission on economic growth as a proposed space exploration theme. The Aldridge Commission identifies the ability of investments in civilian space programs to generate new jobs within current industries and spawn new markets. The contribution that federal space investments make to the nation's economy was not a key factor identified by the Killian Committee.

As a result of its focus on economic growth as a key theme of space exploration, the Aldridge Commission recommended that "NASA's relationship to the private sector, its organizational structure, business culture, and management processes—all largely inherited from the Apollo era—must be decisively transformed to implement the new, multi-decadal space exploration vision." Two of its specific recommendations were that NASA recognize and implement a far larger private industry presence in space operations, with the specific goal of allowing private industry to assume the primary role of providing services to NASA, and that NASA's centers be reconfigured as Federally Funded Research and Development Centers (FFRDCs) to enable innovation, work effectively with the private sector, and stimulate economic development.[37]

FFRDCs are not-for-profit organizations which are financed on a sole-source basis, exclusively or substantially by an agency of the federal government, and not subject to Office of Personnel Management regulations. They operate as private non-profit corporations, although they are subject to certain personnel and budgetary controls imposed by Congress and/or their sponsoring agency. Each FFRDC is administered by either an industrial firm, a university, or a nonprofit institution through a contract with the sponsoring federal agency. FFRDC personnel are not considered federal employees, but rather employees of the organization that manages and operates the center.[38] NASA has not fully adopted the Aldridge Commission recommendations. NASA has 10 centers (see **Table 1**).[39] One, the Jet Propulsion Laboratory (JPL), is already an FFRDC and is managed by the California Institute of Technology.

WHAT IS THE PUBLIC'S ATTITUDE TOWARD SPACE EXPLORATION?

Some editorialists question whether investing in space exploration is relevant today.[40] Others question if NASA has the right priorities.[41] Would the public care if the country's investment in space exploration ended? Does the public believe it would be better to invest in social needs here on Earth rather than space exploration? Does the public support the current prioritization of the nation's space exploration activities?

According to poll data, Americans do not rank space exploration as a high priority for federal government spending. For example, in an April 10, 2007 Harris poll, respondents were given a list of twelve federal government programs and asked to pick two which should be cut "if spending had to be cut." Space programs led the list (51%), followed by welfare programs (28%), defense spending (28%), and farm subsidies (24%).[42] Space exploration was also near the bottom of a University of Chicago National Opinion Research Center survey reported in January 2007 that asked Americans about how they would prioritize federal spending.[43]

Table 1. NASA Centers

Center	Mission Area	Location
Ames Research Center	New Technology Research	Moffett Field, CA
Dryden Flight Research Center	Flight Research	Edwards, CA
Glenn Research Center	Aeropropulsion and Communications Technologies	Cleveland, OH
Goddard Space Flight Center	Earth, the Solar System, and Universe Observations	Greenbelt, MD
Jet Propulsion Laboratory (FFRDC)	Robotic Exploration of the Solar System	Pasadena, CA
Johnson Space Center	Human Space Exploration	TX
Kennedy Space Center	Prepare and Launch Missions Around the Earth and Beyond	Cape Canaveral, FL
Langley Research Center	Aviation and Space Research	Hampton, VA
Marshall Space Flight Center	Space Transportation and Propulsion Technologies	Huntsville, AL
Stennis Space Center	Rocket Propulsion Testing and Remote Sensing	Hancock County, MS, near Slidell

Source: NASA

Note: FFRDC is a federally funded research and development center.

On the other hand, Americans are interested in space exploration. According to a May 2008 Gallup Poll,[44] sponsored by the Coalition for Space Exploration, most Americans (69%) believe that the space program benefits the nation's economy by inspiring young people to consider STEM education, and believe that the benefits of space exploration outweigh the risks of human space flight (68%). The poll also found that most Americans (67%) indicated that they would not be concerned if the United States loses its leadership in space exploration to China, while almost half (47%) of the public surveyed expressed concern regarding the five-year gap between the end of the space shuttle program and the first scheduled launch of the Constellation program. Just over half (52%) of those surveyed in the Gallup Poll said they would support increasing NASA's budget from 0.6% to 1.0% of the federal budget; however, when the public was asked how willing they would be to support an increase in taxes if the money was to go to NASA to help close the budget deficit, more than half (57%) reported they would not be willing.

NASA's Office of Strategic Communication funded several analyses of the public's attitude toward space exploration based on focus groups[45] and a survey,[46] the results of which were presented in June 2007.[47] According to an analysis conducted for NASA, the focus group participants were ambivalent about going to the Moon and Mars and wanted to know why these missions were important. Reasons such as leadership, legacy, and public inspiration were found to be less persuasive, especially for future Moon exploration, than NASA-influenced technologies. Most participants agreed that partnership with other countries would be beneficial, but doubted whether it can be achieved realistically.

In addition, one of the analysis conducted for NASA found that most survey respondents rated NASA-influenced technologies[48] as somewhat or extremely relevant to them. Over 52% of participants said such technologies were a "very strong" reason to go to space. In contrast, the public's response to a mission to send humans to the Moon by the year 2020 was less strong with 15% of respondents very excited and 31% somewhat excited. Results for a mission to send humans to the Mars were similar to those for the Moon.

The public opinion analysis has found that there are generational differences in regard to NASA's proposed activities. For example, NASA's base support came from those who encompass "The Apollo Generation" (45-64 year olds), the majority (79%) of whom support NASA's new space exploration mission, particularly the return to the Moon. By contrast, the majority (64%) of those between 18-24 years of age are uninterested or neutral about a human Moon mission. Those between 25 and 44 years of age are approximately evenly split between those who are interested/excited and those who are either uninterested or neutral. Those over 65 were more likely to be neutral or disinterested in a Moon mission, with those over 75 years of age the least interested of all age groups.[49]

WHAT ARE THE NATION'S PRIORITIES FOR CIVILIAN SPACE EXPLORATION AND ITS IMPLICATIONS FOR FUTURE SPACE POLICY?

Current U.S. civilian space policy is based on a set of fundamental objectives in the Space Act, based on policy discussions that occurred following the launch of Sputnik over 50 years ago. Those objectives are still part of current policy discussions and influence the nation's civilian space policy priorities—both in terms of what actions NASA is authorized to undertake and the degree of appropriations each activity within NASA receives. NASA has active programs that address all its objectives, but many believe that it is being asked to accomplish too much for the available resources.

NASA was last reauthorized in 2008 for FY2009.[50] Thus, the reauthorization of NASA for FY2010 and beyond, along with a new Presidential Administration, may provide an opportunity for Congress to rethink the nation's space policy. The goals of the nation's investment in space exploration may be a key factor in determining the focus of NASA's activities and the degree of funding appropriated for its programs. Congress and outside experts have concerns as to whether the United States can afford to implement President Bush's Vision for Space Exploration without adversely influencing NASA's other programs.[51] Congress may need to make challenging decisions to determine how to reap the most benefit from the nation's civilian space program investment. These decisions might answer questions such as

- What are the priorities among the many reasons for U.S. space exploration? For example, what might be the priority ranking among the previously identified reasons as to why the United States might explore space—knowledge and understanding, discovery, economic growth, national prestige, defense, international relations, and education and workforce development?
- What implications would this prioritization have for NASA's current and future budgets and the balance among its programs? For example, what is the proper balance between human and robotic space activities?
- What influence might the timing of other countries' space exploration activities have on U.S. policy? For example, what would be the impact of the United States, China, or another country, or a commercial organization, establishing the first Moon base or landing on Mars?

New objectives and priorities might help determine NASA's goals. This, in turn, might potentially help Congress determine the most appropriate balance of funding available among NASA's programs during its authorization and appropriation process. For example, if Congress believes that national prestige should be the highest priority, they may choose to emphasize NASA's human exploration activities, such as establishing a Moon base and landing a human on Mars. If they consider scientific knowledge the highest priority, Congress may emphasize unmanned missions and other science-related activities as NASA's major goal. If international relations are a high priority, Congress might encourage other nations to become equal partners in actions related to the International Space Station. If spinoff effects, including the creation of new jobs and markets and its catalytic effect on math and science education, are Congress' priorities, then they may focus NASA's activities on technological development and linking to the needs of business and industry, and expanding its role in science and mathematics education.

Congressional Activities

On October 15, 2008, the NASA Authorization Act of 2008 (P.L. 110-422) was signed into law.[52] This act authorized appropriations for FY2009, and prohibited NASA from taking any steps prior to April 30, 2009, that would preclude the President and Congress from being able to continue to fly the Space Shuttle past 2010. When the law was passed, the Chair of the House Science and Technology Committee stated

> The [Space Shuttle] provision should not be construed as a congressional endorsement of extending the life of the Shuttle program beyond the additional flight added by this bill to deliver the AMS [Alpha Magnetic Spectrometer] to the International Space Station. Rather, it reflects our common belief that the decision of whether or not to extend the Shuttle past its planned 2010 retirement date should be left to the next President and Congress, especially since both of the Presidential candidates have asked for the flexibility to make that decision.[53]

During the 111th Congress, policymakers may discuss another authorization bill for future years, and identify new priorities for civil space exploration.

H.Res. 67 would recognize and commend NASA, the Jet Propulsion Laboratory (JPL), and Cornell University for the success of the Mars Exploration Rovers, Spirit and Opportunity, on the 5th anniversary of the Rovers' successful landing. H.R. 255, the NASA 50th Anniversary Commemorative Coin Act, would require the Secretary of the Treasury to mint coins in commemoration of the 50th anniversary of the establishment of NASA.

The Federal Ocean Acidification Research And Monitoring Act of 2009 (H.R. 14, S. 173) would require that the NASA Administrator ensure that space-based monitoring assets are used in as productive a manner as possible for monitoring of ocean acidification and consistent with a strategic research, and encourage coordination of the Agency's ocean acidification activities with such activities of other nations and international organizations. As passed by the Senate, the same provision is within the Omnibus Public Land Management Act of 2009 (S. 22).

Appendix. Possible U.S. Civilian Space Policy Objectives

Table A-1. Possible U.S. Civilian Space Policy Objectives: Comparison of Selected Extracts from Historical and Current Space Policy Documents

Possible U.S. Civilian Space Policy Objectives	Space Act Objectives (July 1958, as amended)	Eisenhower Administration "Killian Committee" Factors (March 1958)	"Aldridge Commission" Themes (2004)	G.W. Bush Administration "Space Policy" Goals (2006)	NASA Authorization Act (P.L. 110-422) Findings (2009)
Knowledge and Understanding	"(1) The expansion of human knowledge of the Earth and of phenomena in the atmosphere and space."	"Fourth, space technology affords new opportunities for scientific observation and experiment which will add to our knowledge and understanding of the earth, the solar system, and the universe."	"Exploring the Moon, Mars, and beyond is a great journey worthy of a great nation. The impulse to explore the unknown is a human imperative, and a notable part of what animates us as a people. This endeavor presents an opportunity to inspire a new generation of American explorers, scientist, entrepreneurs, and innovators."	"Increase the benefits of civil exploration, scientific discovery, and environmental activities."	"(1) NASA is and should remain a multimission agency with a balanced and robust set of core missions in science, aeronautics, and human space flight and exploration." "(3) Investment in NASA's programs, like investments in other Federal science and technology activities, is an investment in our future." "(9) NASA's scientific research activities have contributed much to the advancement of knowledge, provided societal benefits, and helped train the next generation of scientists and engineers, and those activities should continue to be an important priority."
Discovery	"(2) The improvement of the usefulness, performance, speed, safety, and efficiency of aeronautical and space vehicles." "(3) The development and operation of vehicles capable of carrying instruments, equipment, supplies, and living organisms through space."	"The first of these factors is the compelling urge of man to explore and to discover, the thrust of curiosity that leads men to try to go where no one has gone before. Most of the surface of the Earth has now been explored and men now turn to the exploration of outer space as their next objective."	"Despite the spiritual, emotional, and intellectual appeal of a journey to space—exploration and discovery will perhaps not be sufficient drivers to sustain what will be a long, and at times risky, journey. We must also undertake this mission for pragmatic, but no less compelling reasons, which have everything to do with life here on Earth."	"Implement and sustain an innovative human and robotic exploration program with the objective of extending human presence across the solar system."	"(4) Properly structured, NASA's activities can contribute to an improved quality of life, economic vitality, United States leadership in peaceful cooperation with other nations on challenging undertakings in science and technology, national security, and the advancement of knowledge." "(10) NASA should make a sustained commitment to a robust long-term technology development activity. Such investments represent the critically important ''seed corn'' on which NASA's ability to carry out challenging and productive missions in the future will depend."

Possible U.S. Civilian Space Policy Objectives	Space Act Objectives (July 1958, as amended)	Eisenhower Administration "Killian Committee" Factors (March 1958)	"Aldridge Commission" Themes (2004)	G.W. Bush Administration "Space Policy" Goals (2006)	NASA Authorization Act (P.L. 110-422) Findings (2009)
Economic Growth— Job Creation and New Markets	"(4) The establishment of long-range studies of the potential benefits to be gained from, the opportunities for, and the problems involved in the utilization of aeronautical and space activities for peaceful and scientific purposes."	Not discussed.	"Further space exploration will generate new jobs within current industries and will likely spawn entire new markets involving leading-edge manufacturing and flight support services.... As one impressive labor leader testified to the Commission, 'every dollar spent on space is a dollar spent here on Earth.' This focus is good for jobs, good for the economy, and good for American families."	"Enable a dynamic, globally competitive domestic commercial space sector in order to promote innovation, strengthen U.S. leadership, and protect national, homeland, and economic security."	"(2) Investment in NASA's programs will promote innovation through research and development, and will improve the competitiveness of the United States." "(12) Commercial activities have substantially contributed to the strength of both the United States space program and the national economy, and the development of a healthy and robust United States commercial space sector should continue to be encouraged."
National Prestige	"(5) The preservation of the role of the United States as a leader in aeronautical and space science and technology and in the application thereof to the conduct of peaceful activities within and outside the atmosphere." "(9) The preservation of the United States' preeminent position in aeronautics and space through research and technology development related to associated manufacturing processes."[a]	"Third, there is the factor of national prestige. To be strong and bold in space technology will enhance the prestige of the United States among the peoples of the world and create added confidence in our scientific, technological, industrial, and military strength."	"Exploring the Moon, Mars, and beyond is a great journey worthy of a great nation. The impulse to explore the unknown is a human imperative, and a notable part of what animates us as a people."	"Strengthen the nation's space leadership and ensure that space capabilities are available in time to further U.S. national security, homeland security, and foreign policy objectives."	"(6) NASA should undertake a program of aeronautical research, development, and where appropriate demonstration activities with the overarching goals of— (A) ensuring that the Nation's future air transportation system can handle up to 3 times the current travel demand and incorporate new vehicle types with no degradation in safety or adverse environmental impact on local communities; (B) protecting the environment; (C) promoting the security of the Nation; and (D) retaining the leadership of the United States in global aviation." "(8) Developing United States human space flight capabilities to allow independent American access to the International Space Station, and to explore beyond low Earth orbit, is a strategically important national imperative,

Table A-1. (Continued)

Possible U.S. Civilian Space Policy Objectives	Space Act Objectives (July 1958, as amended)	Eisenhower Administration "Killian Committee" Factors (March 1958)	"Aldridge Commission" Themes (2004)	G.W. Bush Administration "Space Policy" Goals (2006)	NASA Authorization Act (P.L. 110-422) Findings (2009)
					and all prudent steps should thus be taken to bring the Orion Crew Exploration Vehicle and Ares I Crew Launch Vehicle to full operational capability assoon as possible and to ensure the effective development of a United States heavy lift launch capability for missions beyond low Earth orbit. "
Defense	"(6) The making available to agencies directly concerned with national defense of discoveries that have military value or significance, and the furnishing by such agencies, to the civilian agency established to direct and control non-military aeronautical and space activities, of information as to discoveries which have value or significance to that agency."	"Second, there is the defense objective for the development of space technology. We wish to be sure that space is not used to endanger our security. If space is to be used for military purposes, we must be prepared to use space to defend ourselves."	"Much of the United States' current military strength and economic security rests on our technological leadership. Our technological and industrial base must constantly be renewed. Therefore, the United States must continue to lead, especially in those industries that require, and therefore build, technology skills."	"Enable unhindered U.S. operations in and through space to defend our interests there." "Enable a robust science and technology base supporting national security, homeland security, and civil space activities."	"(13) It is in the national interest for the United States to have an export control policy that protects the national security while also enabling the United States aerospace industry to compete effectively in the global market place and the United States to undertake cooperative programs in science and human space flight in an effective and efficient manner."
International Relations	"(7) Cooperation by the United States with other nations and groups of nations in work done pursuant to this act and in the peaceful application of the results thereof."	Not discussed.	"Although the era of Sputnik has given way to an age of international cooperation in space, it remains a competitive frontier ... Other nations, against whom we compete for jobs in the global economy, are also intent on exploring space. If not us, someone else will lead in the exploration, utilization, and ultimately, the commercialization of space, as we sit idly by."	"Encourage international cooperation with foreign nations and/or consortia on space activities that are of mutual benefit and that further the peaceful exploration and use of space, as well as to advance national security, homeland security, and foreign policy objectives."	"(5) NASA should assume a leadership role in a cooperative international Earth observations and research effort to address key research issues associated with climate change and its impacts on the Earth system." "(7) Human and robotic exploration of the solar system will be a significant long-term undertaking of humanity in the 21st century and beyond, and it is in the national interest that the United States should assume a leadership role in a cooperative international exploration initiative."

Table A-1. (Continued)

Possible U.S. Civilian Space Policy Objectives	Space Act Objectives (July 1958, as amended)	Eisenhower Administration "Killian Committee" Factors (March 1958)	"Aldridge Commission" Themes (2004)	G.W. Bush Administration "Space Policy" Goals (2006)	NASA Authorization Act (P.L. 110-422) Findings (2009)
			If not us, someone else will lead in the exploration, utilization, and ultimately, the commercialization of space, as we sit idly by."	objectives."	of the solar system will be a significant long-term undertaking of humanity in the 21 st century and beyond, and it is in the national interest that the United States should assume a leadership role in a cooperative international exploration initiative."
Education and Workforce Development	"(8) The most effective utilization of the scientific and engineering resources of the United States, with close cooperation among all interested agencies of the United States in order to avoid unnecessary duplication of effort, facilities, and equipment."	Not discussed.	"Long-term competitiveness requires a skilled workforce. The space exploration vision can be a catalyst for a much-needed renaissance in math and science education in the United States."	Not discussed.	"(11) NASA, through its pursuit of challenging and relevant activities, can provide an important stimulus to the next generation to pursue careers in science, technology, engineering, and mathematics."

Sources: "Space Act": P.L. 85-568, The National Aeronautics and Space Act, July 29, 1958.This analysis focuses on the objectives section. "Killian Committee": U.S. President (Dwight D. Eisenhower), President's Science Advisory Committee, Introduction to Outer Space, March 26, 1958. p. 2. "Aldridge Commission": U.S. President (George W. Bush), President's Commission on Implementation of United States Space Exploration Policy, A Journey to Inspire, Innovate, and Discovery, June 2004, p. 11. Excerpts are from the section entitled "Why Go?: U.S. President (G.W. Bush), U.S. National Space Policy, August 31, 2006, Excerpts are from section 3, "United States Space Policy Goals." NASA Authorization Act: P.L. 110-422, National Aeronautics and Space Administration Authorization Act of 2008, October 1 5, 2008. This analysis focuses on Section 2, Findings.

Notes: Excerpts are selected to reflect the general tone of text and are not necessarily the only language discussing these issues.

a. The words in italics in the "Space Act" column show the changes made to the objectives since 1958.

End Notes

[1] U.S. President (Dwight D. Eisenhower), President's Science Advisory Committee, *Introduction to Outer Space*, March 26, 1958. p. 1, at http://www.hq.nasa.gov/office/pao/History/monograph10/doc6.pdf.

[2] Ibid., pp. 1-2.

[3] A spinoff is defined by NASA as "A commercialized product incorporating NASA technology or 'know how' which benefits the public." For more information, see NASA, *Spinoff: 50 Years of NASA-Derived Technologies (1958-2008)*, at http://www.sti.nasa.gov/tto/Spinoff2008/index.html.

[4] Sputnik 1 orbited the Earth every 96 minutes until it fell from orbit on January 4, 1958, three months after its launch. Roger D. Launius, "Sputnik and the Origins of the Space Age," at http://history.nasa.gov/sputnik/sputorig.html.

[5] Dwight D. Eisenhower Presidential Center, "Sputnik and the Space Race," at http://www.eisenhower.utexas.edu/dl/Sputnik/Sputnikdocuments.html.

[6] U.S. Congress, House Committee on Science and Technology, *Toward the Endless Frontier: History of the Committee on Science and Technology, 1959-79,* prepared for the Committee by Ken Hechler, committee print, 96th Cong., 2nd sess., H.Prt. 35-120, (Washington: GPO, 1980), pp. 1-28.

[7] P.L. 85-568, The National Aeronautics and Space Act ("Space Act"), July 29, 1958, at http://www.nasa.gov/offices/ ogc/about/space_act1 .html.

[8] DARPA was originally called the Advanced Research Projects Agency (ARPA). It was established by DOD Directive 5105.15 on February 7, 1958, and by Congress in P.L. 85-325 on February 12, 1958. The name was changed from ARPA to DARPA by DoD Directive on March 23, 1972. DARPA was redesignated ARPA by President Bill Clinton in an Administration document on February 22, 1993. ARPA's name was changed back to DARPA by P.L. 104-106 on February 10, 1996. For more information about DARPA and its history, see DARPA, "Defense Advanced Research Project Agency: Technology Transition," January 1997 at http://www.darpa.mil/body/pdf/transition

[9] The appropriation for NSF continued to increase in future years. In 1968, it was almost $500 million. National Science Foundation, An Overview of the First 50 years, at http://www.nsf.gov/about/history/overview-50.jsp.

[10] P.L. 85-864, National Defense Education Act (NDEA), September 2, 1958.

[11] Dwight D. Eisenhower Presidential Center, "Sputnik and the Space Race," at http://www.eisenhower.utexas.edu/dl/Sputnik/Sputnikdocuments.html.

[12] Council on Foreign Relations, Chronology of National Missile Defense Programs, June 1, 2002, at http://www.cfr.org/publication/10443/.

[13] The Project Vanguard booster tests on December 6, 1957 (rose 3 feet, caught fire) and February 5, 1958 (rose 4 miles, exploded) were unsuccessful. A new effort, Project Explorer led by Wernher von Braun, was initiated. Explorer 1 was successful, after two aborted launches, on January 31, 1958. Roger D. Launius, "Sputnik and the Origins of the Space Age," at http://history.nasa.gov/sputnik/sputorig.html.

[14] Sputnik 1 weighed 183 pounds. Sputnik 2 launched on November 3, 1957 weighed 1,120 pounds, carried a dog, and stayed in orbit for almost 200 days. The first satellite to be launched in the American Project Vanguard was planned to be 3.5 pounds. Roger D. Launius, "Sputnik and the Origins of the Space Age," at http://history.nasa.gov/sputnik/ sputorig.html.

[15] Davis, James C., *The Human Story: Our History, From the Stone Age to Today* (New York: Harper Collins, 2004). According to Davis, the statement was made by a Japanese newspaper shortly after the event. Others called it a "technological Pearl Harbor."

[16] Roger D. Launius, "Sputnik and the Origins of the Space Age," at http://history.nasa.gov/sputnik/sputorig.html.

[17] NASA Authorization Act of 2008 (P.L. 110-422).

[18] Space Foundation, The Space Report: Guide to Global Space Activities, 2006, at http://www.thespacereport.org/. For more on the space economy, see Michael D. Griffin, Administrator, National Aeronautics and Space Administration, *The Space Economy*, NASA 50th Anniversary Lecture Series, September 17, 2007, at http://www.nasa.gov/audience/ formedia/speeches/mg_speech_collection_archive_1.html.

[19] Peter Spotts, "Many Contestants in Latest 'Space Race' to the Moon," *Christian Science Monitor*, October 1, 2007.

[20] X-prize Foundation, "Google Sponsors Lunar X PRIZE to Create a Space Race for a New Generation," press release, September 13, 2007, at http://www.xprize.org/lunar/press-release/google-

[21] Virgin Galactic, Overview, at http://www.virgingalactic.com/. NASA has signed a memorandum of understanding with Virgin Galactic to explore the potential for collaborations on the development of space suits, heat shields for spaceships, hybrid rocket motors, and hypersonic vehicles capable of traveling five or more times the speed of sound. See NASA, "NASA, Virgin Galactic to Explore Future Cooperation," press release, February 21, 2007, at http://www.nasa.gov/home/hqnews/2007/feb/HQ_07049_Virgin_Galactic.html.

[22] EADS-Astrium, "Astrium Rockets into Space Tourism," press release, June 13, 2007, at http://www.astrium.eads.net/press-center/press-releases/astrium-rockets-into-space

[23] Jeremy Quittner (ed.), "I Need My Space," *Business Week*, Winter 2007, at http://www.businessweek.com/ magazine/content/07_09/b4023413.htm.

[24] White House, "The Agenda - Defense," webpage at http://www.whitehouse.gov/agenda/defense , accessed February 2, 2009.

[25] U.S. President (G.W. Bush), *U.S. National Space Policy*, August 31, 2006, at http://www.ostp.gov/html/ US%20National%20Space%20Policy.pdf. It replaced the previous space policy that had been in place for 10 years.

[26] U.S. President (G.W. Bush), *President Bush Announces New Vision for Space Exploration Program, Fact Sheet: A Renewed Spirit of Discovery,* January 14, 2004, at http://www.whitehouse.gov/news/releases/2004/01/ 20040114- 1.html.

[27] Twelve U.S. astronauts walked on the Moon between 1969 and 1972. No humans have visited Mars.

[28] U.S. President (G. W. Bush), "A Renewed Spirit of Discovery," document, January 14, 2004, at http://www.whitehouse.gov/space

[29] NASA, *Vision for Space Exploration*, February 2004, at http://www.nasa.gov/mission_pages/exploration/main/ index.html.

[30] CRS Report RS22625, *National Aeronautics and Space Administration: Overview, FY2008 Budget in Brief, and Key Issues for Congress*, by Daniel Morgan and Carl E. Behrens; CRS Report RL33568, *The International Space Station and the Space Shuttle*, by Carl E. Behrens; and CRS Report RS21720, *Space Exploration: Issues Concerning the "Vision for Space Exploration"*, by Marcia S. Smith.

[31] Testimony of Michael D. Griffin, Administrator, National Aeronautics and Space Administration before the Senate Committee on Commerce, Science and Transportation Subcommittee on Space, Aeronautics and Related Sciences, Budget Hearing, U.S. Senate, February 28, 2007, at http://commerceTestimony_MichaelDGriffin_NASA_FY2008PostureStatementFINAL22707.pdf.

[32] National Research Council, Space Studies Board, *An Assessment of Balance in NASA's Science Program*, Washington, DC, 2006, p. 2 http://www.nap.edu/catalog.php?record_id. The NRC is in the midst of a project entitled "Rationale and Goals of the U.S. Civil Space Program." The report from this project is scheduled to be released in July 2009. For more information, see http://www7.nationalacademies.org/ssb/rationale_goals

[33] P.L. 85-568, The National Aeronautics and Space Act, July 29, 1958, at http://www.nasa.gov/offices/ogc/about/ space_act1.html. Since 1958, the objectives have only had two modifications. The clause, "of the Earth and" was added to the first objective by the National Aeronautics and Space Administration Authorization Act, 1985, P.L. 98-361, § I 10(b), 98 Stat. 422, 426 (July 16, 1984). Objective (9) was added by the National Aeronautics and Space Administration Authorization Act, Fiscal Year 1989, P.L. 100-685, § 214, 102 Stat. 4083, 4093 (November 17, 1988). bjective (9) states the following: "The preservation of the United States' preeminent position in aeronautics and space through research and technology development related to associated manufacturing processes."

[34] During the Sputnik era, President Eisenhower's Science Advisory Committee, chaired by George Killian, ("Killian Commission") responded to the fundamental question of why the United States might undertake a national space program in its report *Introduction to Outer Space*. (U.S. President (Dwight D. Eisenhower), President's Science Advisory Committee, *Introduction to Outer Space*, March 26, 1958. p. 2, at http://www.hq.nasa.gov/office/pao/History/ monograph10/doc6.pdf). The President's Science Advisory Committee is analogous to today's President's Council of Advisors on Science and Technology (PCAST).

[35] U.S. President (George W. Bush), President's Commission on Implementation of United States Space Exploration Policy, *A Journey to Inspire, Innovate, and Discovery*, June 2004. Available at http://www.nasa.gov/pdf/ 6073 6main_M2M_report_small.pdf. The commission report is named for its chair, Edward C. "Pete" Aldridge, Jr., and called the "Aldridge Commission" report.

[36] U.S. President (G.W. Bush), *U.S. National Space Policy*, August 31, 2006, at http://www.ostp.gov/html/ US%20National%20Space%20Policy.pdf.

[37] U.S. President (George W. Bush), President's Commission on Implementation of United States Space Exploration Policy, *A Journey to Inspire, Innovate, and Discovery*, June 2004, p. 7, at http://www.nasa.gov/pdf/ 60736main_M2M_report_small.pdf.

[38] See CRS Report RL34454, *Science and Technology Policymaking: A Primer*, by Deborah D. Stine.

[39] For more information, see http://education.nasa.gov/about/nasacenters/index.html.

[40] See, for example, Anne Applebaum, "Mission to Nowhere," *Washington Post*, January 7, 2004, p. A21.

[41] See, for example, Gregg Easterbrook, "How NASA Screwed Up (And Four Ways to Fix It)," *Wired*, May 22, 2007, at http://www.wired.com/science/space *The Economist*, "Spacemen Are from Mars," September 27, 2007, at http://www.economist.com/opinion/displaystory.cfm?story_id

[42] Harris Interactive, "Closing the Budget Deficit: U.S. Adults Strongly Resist Raising Any Taxes Except "Sin Taxes" Or Cutting Major Programs," press release, April 10, 2007, at http://www.harrisinteractive.com/harris_poll/ index.asp?PID=746. The poll was of 2,223 adults surveyed online between March 6 and 14, 2007. This online survey is not based on a probability sample and therefore no theoretical sampling error can be calculated.

[43] University of Chicago, "Americans Want to Spend More on Education, Health," press release, January 10, 2007, at http://www-news.uchicago.edu/releases/07/070110.gss.shtml. The General Social Survey, supported by the National Science Foundation, has been conducted since 1973, and is based on face-to-face interviews of randomly selected people who represent a scientifically accurate cross section of Americans. For the 2006 survey, 2,992 people were interviewed and asked a wide variety of questions in addition to those related to spending priorities.

[44] Coalition for Space Exploration, *2008 Gallup Poll: American Support for Space Exploration is Strong*, at http://www.spacecoalition.com/files/galluppolls/final%20report-june%2008.pdf.

[45] The focus groups were professionally moderated by Dr. Stephen Everett of the Everett Group, Inc., in consultation with ViaNovo. The six focus groups were located in San Diego, Kansas City, and Philadelphia.

[46] The professionally conducted telephone survey was of 1,001 U.S. adults in February 2007. The margin of error was ± 3.2%. The survey was conducted by Dr. Mary Lynne Dittmar of Dittmar Associates, in consultation with ViaNovo.

[47] Robert Hopkins, "Strategic Communications Framework Implementation Plan," powerpoint presentation, NASA, Office of Strategic Communications, June 26, 2007, at http://www.spaceref.com/news/ viewsr .html?pid=24646.

[48] An example of a NASA-influenced technology (commonly called "spinoff") mentioned in the survey that had significant results is a smoke alarm. According to NASA, in the 1970s NASA needed a smoke and fire detector with adjustable sensitivity for Skylab, America's first space station. Honeywell developed the device for NASA and then made it available commercially so that consumers could avoid "nuisance" alarms while cooking. Other devices in the survey were advanced breast cancer imaging, heart defibrillators, weather satellites, remote-controlled robots, global positioning system, cordless tools, satellite radio, and DirecTV. See http://www.sti.nasa.gov/tto/ for more details on NASA's spinoffs. See a list of NASA's top 20 spinoffs in the last five years at http://www.ipp.nasa.gov/ spinoff_top_20a.pdf.

[49] Ibid., p. 9. Robert Hopkins, "Strategic Communications Framework," powerpoint presentation, NASA, Office of Communications Planning, February 2007, at http://images. M. L. Dittmar, *The Market Study for Space Exploration*, (Houston, TX: Dittmar Associates, Inc., 2004), pp. 26-29 (age data) and pp. 8-11 (Executive Summary). M. L. Dittmar, "Engaging the 18-25 Generation: Educational Outreach, Interactive Technologies, and Space". Paper #2006-7303 in Proceedings of AIAA Space 2006, September 19-21, (San Jose, California. Washington, D.C.: AIAA, 2006). Paper and presentation available at http://www.dittmar-associates.com/ Paper_Downloads.htm.

[50] P.L. 109-155, NASA Authorization Act of 2005, December 30, 2005.

[51] See earlier discussion for Senate and House Committee on Appropriations report language; also Lennard Fisk, Chair, Space Studies Board, National Research Council and Thomas M. Donahue Collegiate Professor of Space Science, University of Michigan, *The President's Vision for Space Exploration: Perspectives from a Recent NRC Workshop on National Space Policy*, Testimony before the House Committee on Science, March 10, 2004, at http://science.house.gov/Commdocs/hearings/full04/mar10/fisk.pdf.

[52] For more information, see CRS Report RS22818, *National Aeronautics and Space Administration: Overview, FY2009 Budget, and Issues for Congress*, by Daniel Morgan and Carl E. Behrens.

[53] House Committee on Science and Technology, "House Sends NASA Bill to President's Desk, Reaffirms Commitment to Balanced and Robust Space and Aeronautics Program," press release, September 27, 2008 at http://science.house.gov/press/PRArticle.aspx?NewsID=2309.

Chapter 3

THE INTERNATIONAL SPACE STATION AND THE SPACE SHUTTLE*

Carl E. Behrens

SUMMARY

The International Space Station (ISS) program began in 1993, with Russia joining the United States, Europe, Japan, and Canada. Crews have occupied ISS on a 4-6 month rotating basis since November 2000.

The U.S. Space Shuttle, which first flew in April 1981, has been the major vehicle taking crews and cargo back and forth to ISS, but the shuttle system has encountered difficulties since the *Columbia* disaster in 2003. Russian Soyuz spacecraft are also used to take crews to and from ISS, and Russian Progress spacecraft deliver cargo, but cannot return anything to Earth, since they are not designed to survive reentry into the Earth's atmosphere. A Soyuz is always attached to the station as a lifeboat in case of an emergency.

President Bush, prompted in part by the *Columbia* tragedy, made a major space policy address on January 14, 2004, directing NASA to focus its activities on returning humans to the Moon and someday sending them to Mars. Included in this "Vision for Space Exploration" is a plan to retire the space shuttle in 2010. The President said the United States would fulfill its commitments to its space station partners, and the shuttle *Discovery* made the first post-*Columbia* flight to the ISS in July 2006. Shuttle flights have continued and completion of the space station is scheduled before the shuttle is retired in 2010. Meanwhile NASA has begun development of a new crew launch vehicle, named Ares, and a crew exploration vehicle, named Orion.

NASA programs were funded for FY2008 in Division B of the Consolidated Appropriations Act (P.L. 110-161). The Space Operations program, which includes the space shuttle and the ISS, was funded at $6.734 billion. For FY2009 NASA requested $5.775

---

* This is an edited, reformatted and augmented version of a CRS Report for Congress publication dated March 2009.

billion for these programs, but in the process revised its budgeting to move its overhead costs to a new account called Cross- Agency Support. Under the new system, the FY2008 Space Operations program would have received $5.526 billion, about $250 million less than the FY2009 request. NASA is currently operating under a continuing resolution (Division A of P.L. 110-329), which funded most civilian activities through March 6, 2009. Under the continuing resolution, Space Operations are funded at the $5.526 billion rate appropriated for FY2008.

An FY2009 NASA authorization bill (H.R. 6063) was introduced May 15, 2008. Among the provisions in the one-year authorization bill was a "Sense of the Congress" urging cooperation in the Moon/Mars activities with other nations pursuing human space flight. It also requires that NASA "terminate or suspend any activity of the Agency that, if continued between the date of enactment of this Act and April 30, 2009, would preclude the continued safe and effective flight of the Space Shuttle after fiscal year 2010 if the President inaugurated on January 20, 2009, were to make a determination to delay the Space Shuttle's scheduled retirement." Congress passed the bill September 27, and it was signed by the President October 15 (P.L. 110-422).

Most Recent Developments

With the signing of the Omnibus Appropriations Act for FY2009 (H.R. 1105, P.L. 111-5) on March 11, 2009, the National Aeronautics and Space Administration (NASA) was funded at $17.78 billion for the fiscal year. Of that total, $5.76 billion was allocated to Space Operations, which include the Space Shuttle and the International Space Station. For FY2008, NASA had received $5.53 billion for Space Operations.

As NASA continued to fly Space Shuttle missions to the International Space Station, the agency awaited the announcement of a new administrator to replace Michael Griffin, who resigned in January. On March 11 President Obama said he would appoint a new NASA director soon, and that one of biggest tasks of the new director would be "shaping a mission for NASA that is appropriate for the 21st Century."

The International Space Station (ISS)

NASA launched its first space station, Skylab, in 1973. Three crews were sent to live and work there in 1973-1974. It remained in orbit, unoccupied, until it reentered Earth's atmosphere in July 1979, disintegrating over Australia and the Indian Ocean. Skylab was never intended to be permanently occupied, but the goal of a permanently occupied space station with crews rotating on a regular basis, employing a reusable space transportation system (the space shuttle) was high on NASA's list for the post-Apollo years following the moon landings. Budget constraints forced NASA to choose to build the space shuttle first. The first launch of the shuttle was in April 1981. When NASA declared the shuttle "operational" in 1982, it was ready to initiate the space station program.

In his January 25, 1984 State of the Union address, President Reagan directed NASA to develop a permanently occupied space station within a decade, and to invite other countries to join. On July 20, 1989, the 20th anniversary of the first Apollo landing on the Moon, President George H. W. Bush voiced his support for the space station as the cornerstone of a long-range civilian space program eventually leading to bases on the Moon and Mars. That "Moon/Mars" program, the Space Exploration Initiative, was not greeted with enthusiasm in Congress, primarily due to budget concerns, and ended in FY1993, although the space station program continued.

President Clinton dramatically changed the character of the space station program in 1993 by adding Russia as a partner to this already international endeavor. That decision made the space station part of the U.S. foreign policy agenda to encourage Russia to abide by agreements to stop the proliferation of ballistic missile technology, and to support Russia economically and politically as it transitioned from the Soviet era. The Clinton Administration strongly supported the space station within certain budget limits.

The International Space Station program thus began in 1993, with Russia joining the United States, Europe, Japan, and Canada. An Intergovernmental Agreement (IGA) established three phases of space station cooperation. The IGA is a treaty in all the countries except the United States, where it is an Executive Agreement. It is implemented through Memoranda of Understanding (MOUs) between NASA and its counterpart agencies.

During Phase I (1995-1998), seven U.S. astronauts remained on Russia's space station *Mir* for long duration (several month) missions with Russian cosmonauts, Russian cosmonauts flew on the U.S. space shuttle seven times, and nine space shuttle missions docked with *Mir* to exchange crews and deliver supplies. Repeated system failures and two life-threatening emergencies on *Mir* in 1997 raised questions about whether NASA should leave more astronauts on *Mir*, but NASA decided *Mir* was sufficiently safe to continue the program. (*Mir* was deorbited in 2001.) Phases II and III involve construction of the International Space Station itself, and blend into each other. Phase II began in 1998 and was completed in July 2001; Phase III is underway.

President George W. Bush, prompted in part by the February 2003 space shuttle *Columbia* tragedy, made a major space policy address on January 14, 2004, directing NASA to focus its activities on returning humans to the Moon and eventually sending them to Mars. Included in this "Vision for Space Exploration" was a decision to retire the space shuttle in 2010. The President said the United States would fulfill its commitments to its space station partners.

ISS Design, Cost, Schedule, and Lifetime

Under the original ISS schedule, assembly of the station would have been completed in 2002, with operations at least through 2012. President Bush restructured the space station program in 2001, and left it unclear when assembly would be completed. NASA briefing charts in March 2003 showed space station operations possibly continuing until 2022. Under President Bush's January 2004 "Vision for Space Exploration," however, NASA plans to complete its utilization of ISS in 2016 (though the other partners may continue to use it after that time).

ISS segments have been and continue to be launched into space on U.S. or Russian launch vehicles and assembled in orbit. The space station is composed of a multitude of modules, solar arrays to generate electricity, remote manipulator systems, and other elements.

The U.S. space shuttle has been the major vehicle taking crews and cargo back and forth to ISS, but the shuttle system encountered difficulties after the *Columbia* disaster and did not resume flights until 2006. Russian Soyuz spacecraft are also used to take crews to and from ISS, and Russian Progress spacecraft deliver cargo, but cannot return anything to Earth, since it is not designed to survive reentry into the Earth's atmosphere. A Soyuz is always attached to the station as a lifeboat in case of an emergency.

"Expedition" crews have occupied ISS on a 4-6 month rotating basis since November 2000. Originally the crews had three members (two Russians and one American, or two Americans and one Russian). Crew size was temporarily reduced to two (one American, one Russian) while the U.S. shuttle was grounded in order to reduce resupply requirements. The number of astronauts who can live on the space station is limited in part by how many can be returned to Earth in an emergency by lifeboats docked to the station. Only Russian Soyuz spacecraft are available as lifeboats. Each Soyuz can hold three people, limiting crew size to three if only one Soyuz is attached. The plan is that crew size will grow to six once assembly is completed.

Each Soyuz must be replaced every six months. The replacement missions are called "taxi" flights since the crews bring a new Soyuz up to ISS and bring the old one back to Earth. Therefore, under normal conditions, the long duration Expedition crews are regularly visited by taxi crews, and by the space shuttle bringing up additional ISS segments or exchanging Expedition crews. When the shuttle is unavailable, Expedition crews are taken back and forth on the "taxi" flights.

In order to contract for Soyuz service to the ISS, NASA has needed an exemption from the Iran Nonproliferation Act (INA) (P.L. 106-178), which banned U.S. payments to Russia in connection with the International Space Station (ISS) unless the U.S. President determined that Russia was taking steps to halt proliferation of nuclear weapons and missile technology to Iran. In 2005 Congress amended INA to exempt Soyuz flights to the ISS from the ban through 2011. It also extended the provisions of the INA to Syria and North Korea, and renamed it the Iran, North Korea, and Syria Nonproliferation Act (INKSNA). NASA asked for a legislated extension of this exemption, and waiver authority was extended until July 1, 2016, in the Continuing Appropriations Act of 2009 (P.L. 110-329). (For details see CRS Report RL34477, *Extending NASA's Exemption from the Iran, North Korea, and Syria Nonproliferation Act*, by Carl E. Behrens and Mary Beth Nikitin.)

Space Station Costs

From FY1 994 to FY200 1, the cost estimate for building ISS grew from $17.4 billion to about $25 billion. The $17.4 billion estimate did not include launch costs, operational costs after completion of assembly, civil service costs, or other costs. NASA estimated the program's life-cycle cost (all costs, including funding spent prior to 1993) from FY1985 toFY2012 at $72.3 billion. In 1998, GAO estimated the life-cycle cost at $95.6 billion

(GAO/NSIAD-98-147). More recent, comparable, life-cycle estimates are not available from NASA or GAO.

As costs continued to rise, Congress voted to legislate a $25 billion cap on development of the ISS program, plus $17.7 billion for associated shuttle launches, in the FY2000-FY2002 NASA authorization act (P.L. 106-391). In January 2001, however, NASA announced that the cost would be over $30 billion, 72% above the 1993 estimate, and $5 billion above the legislated cap. NASA explained that program managers had underestimated the complexity of building and operating the station. The Bush Administration signaled it supported the legislated cap, would not provide additional funds, and NASA would have to find what it needed from within its Human Space Flight account.

"Core Complete" Configuration

In February 2001, the Bush Administration announced it would cancel or defer some ISS hardware to stay within the cap and control space station costs. The decision truncated construction of the space station at a stage the Administration called "core complete." In 2001, the space station program office at Johnson Space Center (JSC) estimated that it would cost $8.3 billion from FY2002 toFY2006 to build the core complete configuration, described at that time as all the U.S. hardware planned for launch through "Node 2," plus the launch of laboratories being built by Europe and Japan. NASA subsequently began distinguishing between "U.S. Core Complete" (the launches through Node 2, which, prior to the *Columbia* tragedy, was scheduled for February 2004) and "International Partner (IP) Core Complete" which included the addition of European and Japanese laboratory modules (then anticipated in 2008).

The new policy was followed by President Bush's January 2004 "Vision for Space Exploration," which directs that U.S. research on ISS be restricted only to that which supports the Vision. A new research plan, incorporating the President's Vision, was issued by NASA in June 2006, as mandated by the 2005 NASA authorization act (P.L. 109-155).[1]

At a January 2005 Heads of Agency meeting, the partners endorsed a final configuration of ISS, but NASA subsequently announced changes to it. The agency now plans to conduct only 16 (instead of 28) shuttle launches to the ISS, all before the end of FY2010 (September 30, 2010), and has dropped plans to launch the centrifuge and its accommodation module, and Russia's Science Power Platform. The agency plans to meet with the other ISS partners to discuss these changes.

The changes to the ISS are largely due to the new direction NASA is taking in response to the Vision for Space Exploration. The Vision calls for development of a Crew Exploration Vehicle, now named Orion, to take astronauts to and from the Moon, and a Crew Launch Vehicle, now named Ares I. Orion also can take them to and from the ISS, and NASA Administrator Griffin stated at a September 19, 2005 press conference that Orion would be used to take crews to and from the ISS, and to serve as a lifeboat for them. If Orion is built as announced, it would fulfill the U.S. commitment to build a crew return capability, and allow the ISS crew size to increase to its originally planned complement of seven. An Earth-orbit capability is planned by 2014 (although NASA now considers early 2015 more likely) with the ability to take astronauts to and from the Moon following no later than 2020.

THE SPACE SHUTTLE

The Space Transportation System (STS)—the Space Shuttle—is a partially reusable launch vehicle and is the sole U.S. means for launching humans into orbit. It consists of an airplane-like Orbiter, with two Solid Rocket Boosters (SRBs) on each side, and a large, cylindrical External Tank (ET) that carries fuel for the Orbiter's main engines. The Orbiters and SRBs are reused; the ET is not. NASA has three remaining spaceflight-worthy Orbiters: *Discovery*, *Atlantis*, and *Endeavour*.

Table 1. U.S. Space Station Funding (in $ millions)

Fiscal Year	Request	Appropriated
1985	150	150
1986	230	205
1987	410	410
1988	767	425
1989	967	900
1990	2,050	1,750
1991	2,430	1,900
1992	2,029	2,029
1993	2,250	2,100
1994	2,106	2,106
1995	2,113	2,113
1996	2,115	2,144
1997	2,149	2,149
1998	2,121	2,441[a]
1999	2,270	2,270
2000	2,483	2,323
2001	2,115	2,115
2002	2,114	2,093
2003	1,839	1,810
2004[b]	2,285	2,085
2005	2,412	2,058
2006	1,995	1,972
2007	1,894	NA[c]
2008	1,894	2,209
2009	2,060[d]	2,060

Note: These numbers reflect NASA's figures for "the space station program." Over the years, what is included in that definition has changed. In recent years, funding for ISS research has been located in a different account from ISS development funding. The figures here represent the ISS development and ISS research request and appropriations to the maximum extent possible.

a. NASA's FY1999 budget documents showed $2.50 1 billion in the expectation Congress would approve additional transfer requests, but it did not.

b. Reflects shift to full cost accounting.

c. Space shuttle funding was not specified in P.L. 110-5.

d. Reflects change in budgeting for overhead costs.

The *Challenger* and *Columbia* Tragedies

More than 100 shuttle launches have taken place since April 1981. Two ended in tragedy, each killing seven astronauts. In 1986, the space shuttle *Challenger* exploded 73 seconds after launch because of the failure of a seal (an O-ring) between two segments of an SRB. In 2003, the space shuttle *Columbia* disintegrated as it returned to Earth after 16 days in orbit (see CRS Report RS2 1408, *NASA's Space Shuttle Program: The Columbia Tragedy, the Discovery Mission, and the Future of the Shuttle*, by Marcia S. Smith). A hole in *Columbia's* left wing, caused during launch by a piece of foam insulation that detached from the ET, allowed hot gases to enter the wing during reentry, deforming it and causing the shuttle to break up. The *Columbia* Accident Investigation Board (CAIB) found that the tragedy was caused by technical and organizational failures, and made 29 recommendations, 15 of which it said should be completed before the shuttle returned to flight.[2]

Sean O'Keefe, NASA's Administrator from December 2001-February 2005, said NASA would comply with the CAIB recommendations.

Return to Flight (RTF)

NASA launched the space shuttle *Discovery* on the first of two "Return to Flight" (RTF) missions—STS-114—on July 26, 2005, and it successfully landed on August 9. On July 27, however, NASA announced that a piece of foam had detached from STS-114's ET during launch, similar to what happened to *Columbia*. Cameras and other sensors on *Discovery* and on the International Space Station—to which *Discovery* was docked for much of its mission—imaged the Orbiter and determined that it was not damaged, but further shuttle launches were suspended. Meanwhile, the images revealed that two "gapfillers"—ceramic coated fabric placed between thermal protection tiles—were protruding on the belly of the Orbiter that could have affected aerodynamic heating during reentry. One of the *Discovery* astronauts removed them during a space walk. The second RTF mission—STS-121—was scheduled for September 2005, but deferred.

STS-121 launched on July 4, 2006, and returned safely to Earth on July 17. The shuttle *Atlantis* launched September 9 on STS 115, during which construction of the International Space Station was resumed.

Current plans for the shuttle include nine more flights to complete the ISS before the shuttle is permanently grounded in 2010. Also planned is another flight to service the Hubble Space Telescope. Following the Columbia disaster, then-Administrator Sean O'Keefe had cancelled the Hubble servicing mission, partly on the grounds that shuttle astronauts would not be able to reach the ISS as a haven in case the shuttle was unable to return to earth. The decision was put under review by the new Administrator, Michael Griffin, and on October 31, 2006, he announced that the Hubble mission would be undertaken in 2008. The servicing would extend the life of the telescope through 2013. To deal with emergencies, NASA planned to prepare a "launch on need" mission with a second shuttle ready to launch on a rescue mission if the first was found defective during the servicing mission.

The launch was in final stages of preparation when a major data handling unit already in place in the Hubble telescope failed in late September 2008. A backup unit in the telescope was activated, but the service mission was revised to include carrying a second data handling unit, which had been stored on earth, to replace the failed one. NASA determined that assessing and preparing the second unit for installation and service would delay the mission until May or June of 2009.

The Shuttle's Future

NASA attempted unsuccessfully for many years to develop a "second generation" reusable launch vehicle (RLV) to replace the shuttle. In 2002 NASA indicated the shuttle would continue flying until at least 2015, and perhaps 2020 or beyond. The *Columbia* tragedy, and President Bush's 2004 Vision for Space Exploration—to return astronauts to the Moon by 2020 and someday send them to Mars—forced NASA to revise that plan.

The President's Vision calls for the shuttle program, which absorbs approximately 25% of NASA's annual budget, to be terminated in 2010. A primary motivation is to make that funding available to implement other aspects of the Vision, although there also is concern about shuttle safety. Congress has been debating the Vision, including its impact on the shuttle and on U.S. human access to space. Some Members wanted to terminate the shuttle earlier than 2010 because they feel it is too risky and/or that the funds should be spent on accelerating the Vision. Others want to retain the shuttle at least until a new spacecraft, the Crew Exploration Vehicle (CEV), is available to take astronauts to and from the ISS. The CEV is now planned for 2015 at the earliest, leaving a multi-year gap during which U.S. astronauts would have to rely on Russia for access to the ISS.

The 2008 NASA Authorization Act (P.L. 110-422) included a provision requiring NASA to "terminate or suspend any activity of the Agency that, if continued between the date of enactment of this Act and April 30, 2009, would preclude the continued safe and effective flight of the Space Shuttle after fiscal year 2010 if the President inaugurated on January 20, 2009, were to make a determination to delay the Space Shuttle's scheduled retirement." (Sec. 61 1d.)

Shuttle Budget

Funding for the shuttle for FY2008 was $3.98 1 billion. For FY2009, NASA requested $2.982 billion for the shuttle, but that amount reflects NASA's new system for funding program overhead costs, which created a new Cross-Agency Support account. By the new accounting system, the comparable shuttle funding for FY2008 was $3.267 billion. The omnibus appropriations bill (P.L. 111-8) appropriated the requested $2.982 billion. (For details on the NASA budget, see CRS Report RS228 18, *National Aeronautics and Space Administration: Overview, FY2009 Budget, and Issues for Congress*, by Daniel Morgan and Carl E. Behrens.)

Issues for Congress

In passing the 2005 NASA authorization act (P.L. 109-105), Congress basically agreed with the President's plan for directing NASA's attention to a return to the Moon and manned missions to Mars. Included in the Moon-Mars "Vision" is the plan to end flights of the Space Shuttle in 2010, and restriction of U.S. experiments on the ISS mostly to those that forward the goal Moon-Mars goal. A number of critical questions remain, however.

- Adequacy of funding is the chief question raised about NASA's activities. In presenting the Moon-Mars vision, the President did not request significantly increased money for NASA, despite chronic indications that the missions it was already charged with were underfunded. NASA has responded to the new mission by cutting back funding for its other activities, primarily in scientific research and aeronautics.
- Although *Discovery's* "Return to Flight" mission of July 2006 was a success, the ability of the shuttle fleet to carry out enough flights to complete construction of the ISS by 2010 is still in question. With a history of more than a hundred successful missions, it might be assumed that another 15 or so would be considered more or less routine, but instead, each launch is still a major and risky event. The great complexity of the vehicle and the extreme environment in which it operates require constant attention to possible accidents and malfunctions, many of which must be addressed on an ad hoc basis.
- The future role of the ISS is also unclear. Assuming that enough shuttle flights are made to carry out "core completion" of the station by 2010, it is not clear what will be done with the ISS after that. In particular, there will be a gap of several years between retirement of the shuttle in 2010 and beginning of flight of the Crew Exploration Vehicle, to be designed for the return to the moon but able to serve as a vehicle to reach the ISS.

End Notes

[1] The National Aeronautics and Space Administration (NASA).Research and Utilization Plan for the International Space Station (ISS) A Report to the Committee on Science of the United States House of Representatives and the Committee on Commerce, Science, and Transportation of the United States Senate. June 2006. http://www.exploration.nasa.gov/documents/reports/NASA_Research_and_Utilization_Plan_for_the_ISS.pdf

[2] National Aeronautics and Space Administration. Columbia Accident Investigation Board Report. August 2003. See CRS Report RS2 1606, NASA's Space Shuttle Columbia: Synopsis of the Report of the Columbia Accident Investigation Board, by Marcia S. Smith.

Chapter 4

TESTIMONY OF NORMAN R. AUGUSTINE, BEFORE THE COMMITTEE ON SCIENCE AND TECHNOLOGY, U.S. HOUSE OF REPRESENTATIVES - HEARING ON "OPTIONS AND ISSUES FOR NASA'S HUMAN SPACE FLIGHT PROGRAM"*

Mr. Chairman and Members of the Committee, thank you for this opportunity to share with you the principal findings of the Review of the U.S. Human Spaceflight Plans Committee. I will speak on behalf of the members of our committee and will do my best to reflect our consensus views. As you are aware, our final report has not yet been published; however, our decision-making deliberations were all conducted in public under FACA rules so I believe what I have to say will come as no surprise to anyone.

First, I would like to acknowledge the contributions and extraordinary effort of each of my colleagues on the Committee. Their names and primary affiliations are appended to this statement. I would also like to acknowledge the forthright, responsive and highly professional support we received from NASA as well as from the Aerospace Corporation, the latter of which the Committee employed to provide independent technical and cost assessments.

The Committee was comprised of ten members having highly diverse backgrounds. It included astronauts, scientists, engineers, former presidential appointees, business executives, educators and an Air Force retired General Officer—each with considerable space experience. Due to the exigencies of the budget process we were asked to complete our task in ninety days—which we did, with the exception of finalizing and printing our report. The latter will be available soon.

Our assigned task was to identify alternative courses that the U.S. might pursue in the area of human spaceflight. One such alternative, of course, is to continue the present program. As noted in the Committee's report, changes to ongoing programs are generally warranted only for compelling reasons. Each alternative identified by the Committee is accompanied by a discussion of its strengths and weaknesses.

---

* This is an edited, reformatted and augmented edition of statement/testimony given before the House of Representatives on Septmber 15, 2009.

It was agreed that at least two of the alternatives would be compatible with the FY '10 budget plan extended through FY '20. We were also asked to examine the current plans for the Space Shuttle and International Space Station and, if appropriate offer alternatives thereto. It is important to note that we specifically were not asked to make a *recommendation* as to a future course of action. That decision is, of course, the purview of the President and the Congress.

Before addressing destinations and architectures the Committee sought to identify appropriate goals for human spaceflight. There are many possibilities that can be cited: strengthening the economy, conducting science, repairing and upgrading spacecraft on orbit, promoting international ties, protecting against asteroids and comets, encouraging science education, and more. It is, however, the Committee's view that although each of these benefits is important in its own right, none can, by itself, justify the cost and risk of human spaceflight. Rather, the *raison d'etre* for such activity must, and in our view can, be founded upon charting a course for the expansion of civilization into the solar system. In so doing, one derives the leadership benefits of being among the world's space-faring nations—a nation that is committed to exploration, seeking knowledge, advancing engineering capabilities, inspiring its citizens, and motivating its young people to consider careers in science and engineering. To a not inconsiderable degree it is intangibles that justify the human spaceflight program, intangibles such as those that today help maintain America as a leader among the world's nations. The Apollo Program is an appropriate example.

In carrying out the charge to identify options the Committee narrowed over 3,000 theoretically possible outcomes to a set of five alternative integrated space programs. These can be thought of as representative families, since one can interchange certain elements among the individual alternatives. The Committee's attempt was, of course, to keep the number of nominal options to a manageable size.

The alternatives offered include the ongoing program, Constellation—that is, the Program of Record and the Budget of Record—and four primary alternatives, some having derivatives or "sub-cases."

Two of the five alternatives were in fact constrained to the current budget profile for human spaceflight. The first of these was the Program of Record; that is, today's program, modified to fly-out the Shuttle in 2011 rather than 2010 and including sufficient funds to de-orbit the International Space Station (ISS) in 2016 according to plan. Under this existing approach the Ares I launch vehicle and Orion capsule are unlikely to become available until after the ISS has been de-orbited. The heavy-lift vehicle, Ares V, would, in our judgment, become available in the late 2020's; however, there are inadequate funds to develop the exploration systems the Ares V is intended to support. The Committee concludes that this is not an executable option due to the incompatibility of the budget plan and the program plan.

The Committee's review noted that the Constellation Program has encountered technical difficulties of the type not unexpected of undertakings of this magnitude— problems which, given adequate funds and engineering attention, should be solvable. This was not, however, a significant factor in the overall conclusion with respect to the viability of the Program of Record.

The second of the options, also constrained to the current budget profile, flies-out the Shuttle in FY '11, but extends the use of the International Space Station for five years, to 2020. This option includes a robust technology development program— something the Committee believes has been lacking at NASA in recent years—and relies on commercial firms to launch cargo and crews to the ISS as soon as demonstrated capabilities exist. It includes development of a somewhat less capable version of the Ares V, known as the Ares V (Lite). This option is deemed capable of execution but cannot provide the space-borne hardware required to support a viable exploration program. In fact, the Committee could find no program within the current budget profile that would enable a viable exploration effort.

Given these findings, the Committee examined three options that exceeded the present budget plan. The most defensible funding profile, purely from a program execution standpoint, is one that linearly increases to $3B above the FY '10 guidance by FY '14 and then increases by an estimated annual inflation rate of 2.4 percent.

The first of these budgetarily less constrained options is termed the Baseline Case. It is the present Program of Record with funds added to extend Shuttle operations into 2011 and, as now provided in the budget plan, to de-orbit the ISS in 2016. This program would permit a human return to the moon in the mid '20's and begin laying the groundwork for a flight to Mars.

The second of the budgetarily less constrained cases is actually a family of variants that would extend ISS operations to 2020, provide funds for its de-orbit, and fund a strong technology program in support of ISS utilization and an eventual human landing on Mars. It would use commercial launch services for new access to low-earth orbit. There are, however, significant differences between the two variants under this option. The first of these variants would develop the Ares V (Lite) to support a human lunar landing in the mid 2020's—after which focus would turn to a human Mars landing. The second variant would extend the use of the (recertified) Space Shuttle to 2015 and be accompanied by the development of a Shuttle Directly-Derived heavy-lift vehicle in place of the Ares family—with the eventual possibility of in-orbit refueling. This is the only practicable option the Committee could find to close the at least five-year gap during which the U.S. will, as currently planned, rely upon Russian launch services to lift U.S. astronauts to the International Space Station.

The third budgetarily less constrained case follows a rather different path of exploration from that heretofore pursued by the U.S. The Committee terms this option the "Flexible Path" and defines it as achieving periodic milestones prior to a Moon or Mars landing. These initial accomplishments could include a lunar fly-by, a Mars fly-by, a visit to a Lagrange point, an asteroid rendezvous, and possible landings on the moons of Mars, Phobos and Demos.

In summary, with the existing budget plan it would be reasonable to extend the use of the ISS for five years and to conduct a robust technology development program. The Committee concludes that no rational exploratory program can be funded under the existing funding constraint and that plans for America's space exploration program would de facto be halted and human operations limited to low earth orbit.

With the less constrained budget option, requiring approximately $3B per year in additional funding, a sound exploration program could be conducted. The reason for this seemingly "dead space" between the two budget options is, simplistically stated, that for sixty percent of the needed funds, one cannot go sixty percent of the way to Mars.

Each of the implementable options that was identified has its own set of benefits and liabilities that the Committee has sought to address. The findings of this effort are discussed in the Summary Report. The assessment gives overarching priority to safety and, as is noted in the Summary Report, the Committee believes considerable caution is in order when comparing analytical results in this area with flight results. Similarly, the Committee has sought to be conservative in its cost estimation practices—reflecting dissatisfaction with historical experience on a broad spectrum of programs. Finally, in defining a "Program of Record" the Committee has relied upon NASA's current program plan and the President's budget profile, the latter as provided by the Office of Management and Budget.

In the opinion of this Committee, as well as that of most of the persons with whom the Committee has had contact, NASA has for too long sought to operate in an environment where means do not match ends. In the unforgiving arena of human spaceflight this is a particularly hazardous policy to embrace.

The Committee also notes that NASA has become a mature organization, an organization long protected from restructuring Centers, facilities and personnel cadres. The consequence is an organization with high fixed costs of the type that make budgetary options highly limited. While NASA is unarguably the finest space organization in the world and a great national asset, it is overdue for a thorough management assessment of the type the aerospace industry underwent at the end of the Cold War.

The Committee's report will contain more detailed information that it hopes will prove of value. On behalf of my colleagues, I thank you for the trust that has been placed in us to review a pursuit which for decades has come to be a symbol of America's leadership.

U.S. Human Space Flight Plans Committee Members

Mr. Norman R. Augustine
Retired Chairman & CEO
Lockheed Martin Corporation

Dr. Wanda M. Austin
President & Chief Executive Officer
The Aerospace Corporation

Mr. Bohdan I. Bejmuk
Chair, NASA Constellation Standing Review Board

Dr. Leroy Chiao
Former Astronaut, Former International Space
Station Commander and Engineering Consultant

Dr. Christopher F. Chyba
Professor of Astrophysics Sciences and International Affairs
Princeton University

Dr. Edward F. Crawley
Ford Professor of Engineering
Massachusetts Institute of Technology

Mr. Jeffrey K. Greason
Co-founder & Chief Executive Officer

XCOR Aerospace

Dr. Charles F. Kennel
Director and Professor Emeritus
Scripps Institution of Oceanography
University of California, San Diego

General Lester Lyles
United States Air Force (Retired)

Dr. Sally Ride
President & Chief Executive Officer
Imaginary Lines

Norman R. Augustine was raised in Colorado and attended Princeton University where he graduated with a BSE in Aeronautical Engineering, magna cum laude, and an MSE. He was elected to Phi Beta Kappa, Tau Beta Pi and Sigma Xi.

In 1958 he joined the Douglas Aircraft Company in California where he worked as a Research Engineer, Program Manager and Chief Engineer. Beginning in 1965, he served in the Office of the Secretary of Defense as Assistant Director of Defense Research and Engineering. He joined LTV Missiles and Space Company in 1970, serving as Vice President, Advanced Programs and Marketing. In 1973 he returned to the government as Assistant Secretary of the Army and in 1975 became Under Secretary of the Army, and later Acting Secretary of the Army. Joining Martin Marietta Corporation in 1977 as Vice President of Technical Operations, he was elected as CEO in 1987 and chairman in 1988, having previously been President and COO. He served as president of Lockheed Martin Corporation upon the formation of that company in 1995, and became CEO later that year. He retired as chairman and CEO of Lockheed Martin in August 1997, at which time he became a Lecturer with the Rank of Professor on the faculty of Princeton University where he served until July 1999.

Mr. Augustine was Chairman and Principal Officer of the American Red Cross for nine years, Chairman of the Council of the National Academy of Engineering, President and Chairman of the Association of the United States Army, Chairman of the Aerospace Industries Association, and Chairman of the Defense Science Board. He is a former President of the American Institute of Aeronautics and Astronautics and the Boy Scouts of America. He is a current or former member of the Board of Directors of ConocoPhillips, Black & Decker, Proctor & Gamble and Lockheed Martin, and was a member of the Board of Trustees of Colonial Williamsburg. He is a Regent of the University System of Maryland, Trustee Emeritus of Johns Hopkins and a former member of the Board of Trustees of Princeton and MIT. He is a member of the Advisory Board to the Department of Homeland Security, was a member of the Hart/Rudman Commission on National Security, and served for 16 years on the President's Council of Advisors on Science and Technology. He is a member of the American Philosophical Society and the Council on Foreign Affairs, and is a Fellow of the National Academy of Arts and Sciences and the Explorers Club.

Mr. Augustine has been presented the National Medal of Technology by the President of the United States and received the Joint Chiefs of Staff Distinguished Public Service Award.

He has five times received the Department of Defense's highest civilian decoration, the Distinguished Service Medal. He is co-author of *The Defense Revolution* and *Shakespeare In Charge* and author of *Augustine's Laws* and *Augustine's Travels*. He holds 23 honorary degrees and was selected by Who's Who in America and the Library of Congress as one of "Fifty Great Americans" on the occasion of Who's Who's fiftieth anniversary. He has traveled in over 100 countries and stood on both the North and South Poles of the earth.

Chapter 5

STATEMENT OF VADM JOSEPH W. DYER, USN (RETIRED), CHAIRMAN, NATIONAL AERONAUTICS AND SPACE ADMINISTRATION'S AEROSPACE SAFETY ADVISORY PANEL, BEFORE THE COMMITTEE ON SCIENCE AND TECHNOLOGY, U.S. HOUSE OF REPRESENTATIVES - HEARING ON "OPTIONS AND ISSUES FOR NASA'S HUMAN SPACE FLIGHT PROGRAM" *

Chairman Gordon, Ranking Member Hall and distinguished Members, thank you for the opportunity to appear before you today. As requested, I would like to present the perspective of the NASA Aerospace Safety Advisory Panel with regard to the Options and Issues for NASA's Human Space Flight Program.

The Aerospace Safety Advisory Panel (ASAP) was originally established under Section 6 of the NASA Authorization Act, 1968 (42 U.S.C. § 2477). In 2005, the ASAP authority was modified under Section 106 of the NASA Authorization Act of 2005 (P.L. 109-155).

The ASAP's charge is, among other things, to advise the NASA Administrator and the Congress with respect to the hazards of proposed or existing facilities and proposed operations with respect to the adequacy of proposed or existing safety standards, and with respect to management and culture related to safety.

My goal this afternoon is to share with the Committee much of the same information I shared with the Review of U.S. Human Space Flight Plans Committee on 14 July of this year. I shall restrict my remarks to safety and safety-related opportunities and issues.

In general, the ASAP is both respectful and appreciative of the summary report released by the Review of U.S. Human Space Flight Plans Committee. They quickly conducted a broad and far reaching review of current plans and potential alternatives. The ASAP does

---

* This is an edited, reformatted and augmented edition of statement/testimony given before the House of Representatives on Septmber 15, 2009.

believe the tempo and time available prevented the thorough consideration of risks and safety challenges that we would have liked to have seen.

We note that the Review of U.S. Human Space Flight Plans Committee summary report compares current plans for the Constellation program with a number of conceptual alternatives. Here, we offer a word of caution -- PowerPoint presentations addressing future programs will always out shine current programs of record. Why is that the case? It is because current programs have garnered the professional peer and public review during the accomplishment of real work. Technical challenges will have been discovered, cost stress will have been revealed, and the reality of conducting high risk business in an unforgiving environment will have been highlighted and publicized. Future concepts do not yet have the benefit of this reality testing. This experience led to one of the ASAP's prime recommendations presented to the Review of U.S. Human Space Flight Plans Committee. Specifically, the ASAP believes that if Constellation is not the optimum answer, then *any other new design must be substantially superior to justify starting over.*

"Starting over" would surely and substantially extend the gap in the Nation's ability to transport humans into space. As it is directly related, I want to share the ASAP's strongly held position regarding the Shuttle: *ASAP does not support extending the shuttle beyond the current manifest.* The substantiation of this recommendation is addressed in the Aerospace Safety Advisory Panel 2008 Annual Report which I respectfully request be included in the hearing record.

The ASAP's 2008 Annual Report also addresses the NASA Commercial Orbital Transportation Services (COTS) Project. The Panel noted NASA needs to *take a more aggressive role articulating human rating requirements for the COTS Project.* COTS vehicles currently are not subject to the Human-Rating Requirements (HRR) standards and are not proven to be appropriate to transport NASA personnel. There is no evidence that the COTS vehicles will be completed in time to minimize the gap between n Shuttle and the follow-on program. Additionally, we note that NASA, and at least one of the COTS funded partners, hold widely divergent views as to what is required for human-rating.

An area where the ASAP and the Review of U.S. Human Space Flight Plans Committee are in strongest possible agreement is with regard to budget. The ASAP has noted the need for NASA and the Congress to address an *imperative to achieve better harmony among requirements, resources and acquisition strategy.* The inevitable pressure to shortcut good process in the face of a budget shortfall is THE most damaging infliction upon a proper safety culture and the conduct of good design.

Making better use of robots is another area where the Review of U.S. Human Space Flight Plans Committee and the ASAP have made similar recommendations. The ASAP believes unmanned systems – both stand alone and integrated with astronauts – offer potential to reach farther and to improve safety. The ASAP has highlighted *the role of unmanned systems in support of human exploration in the next decade requires clarification by NASA.* Historically, NASA robots have been embraced mostly by the scientific community and to a much lesser extent by human space flight programs. NASA will be well served to better develop the process by which manned and unmanned systems are integrated. Undertakings as diverse as construction and mining demand coordinated manned and unmanned systems design.

Given good direction, consistency of purpose, and sufficient resources, *Constellation, or an alternative program, offers a one-time opportunity for safety to be better hardwired into overall NASA processes*. Experience shows one of the best ways for a large organization to advance the state of art of its processes is to institutionalize procedures developed by a major new program that is highly motivated and staffed with the best and brightest. We would have liked for the Review of U.S. Human Space Flight Plans Committee to have more strongly highlighted this point as well.

Lastly, the ASAP would like to champion a broader discussion of risk. Lives will be lost in future human exploration of space. We are lucky to have brave men and women willing to undertake exploration in support of mankind even in the face of these risks. We believe there is need for greater dialogue about risk and that *NASA, the White House and the Congress must all shoulder the burden of risk and the necessity of being more transparent with the citizens of our country regarding that risk.*

Chairman Gordon, Ranking Member Hall, and distinguished Members, in closing I would like to note that the new NASA Administrator, Charlie Bolden, has been a member of the ASAP for the last several years. We know him very well and take strong comfort in his ability to lead the Agency during these challenging times. I thank you again for the opportunity to appear today.

In: Future of U.S. Human Spaceflight: Background and Issues ISBN: 978-1-61668-774-8
Ed: Derek A. Warren and Bridget D. Conway

Chapter 6

TESTIMONY OF MICHAEL D. GRIFFIN, EMINENT SCHOLAR AND PROFESSOR, MECHANICAL AND AEROSPACE ENGINEERING, UNIVERSITY OF ALABAMA, HUNTSVILLE, BEFORE THE COMMITTEE ON SCIENCE AND TECHNOLOGY, U.S. HOUSE OF REPRESENTATIVES - HEARING ON "OPTIONS AND ISSUES FOR NASA'S HUMAN SPACE FLIGHT PROGRAM"*

Thank you for inviting me to appear before this committee to discuss our nation's plans for human spaceflight, and the findings of a highly respected Commission chartered to review those plans. For this Hearing, I have been asked to provide my perspective on the Commission's summary report, especially as it relates to NASA's Constellation Program and the issues that Congress needs to consider as it deliberates the future of U.S. human spaceflight. I am honored to have been asked.

I will begin by acknowledging my own gratitude to the Commission for highlighting, front and center, many issues with which I grappled for four years during my term as Administrator. The Commission has offered many observations with which I most strongly agree. Among these are:

- the reaffirmation of the fundamental strategic goal of human expansion outward into the solar system;
- the explicit enunciation of both intangible and concrete reasons— what I once labeled 'real' vs. 'acceptable' reasons -- for human expansion into space;
- the absolute criticality of stable policy direction to the success of such an effort, and the resources to implement that direction, across presidential administrations;

* This is an edited, reformatted and augmented edition of statement/testimony given before the House of Representatives on Septmber 15, 2009.

- the recognition of the impact of substantial, consistent, long-term real-dollar budget cuts at NASA (more than 20% in the last 15 years),
- the plain acknowledgement that more money is required if worthy goals are to be attained, and that without such funding, worthy goals in human spaceflight beyond the International Space Station (ISS) will not be achieved;
- the identification of a specific amount for a proposed increase, $3 billion annually, rather than merely stating a requirement for "more money";
- the value of U.S. leadership in a program of human expansion into space, while still embracing strategically critical contributions by international partners;
- the distinct but complementary natures of scientific discovery and human spaceflight in the expansion of the human frontier;
- the requirement to implement this expansion with a transportation infrastructure designed to last decades and enable numerous destinations;
- the importance of heavy-lift launch systems to that implementation scheme;
- support for the continuation of ISS operations through 2020 (and I would add "at least 2020");
- the need for and benefit of a focused effort in technology development and maturation as part of the overall space exploration enterprise.

The Commission is to be further congratulated for its forthright willingness to engage some ofthe more contentious questions in what has been a long-term but still unsettled policy discussion. There are a number of "hot-buttons" in the report that have been and will continue to be debated passionately until finally settled by decisions and actions. Among these questions are:

- whether or not there is a need for independent U.S. government human access to space, and if not, the identification of those entities upon which we are willing to depend for such access;
- whether or not it is in the larger interests of the United States to invite international partnerships in regard to capabilities which are on the so-called "critical path" to a desired common goal;
- the degree to and roles in which the U.S. government should foster the development, and embrace the capabilities, of "commercial space" in the furtherance of national goals;
- the proper role of NASA in guiding the human expansion into space, and in particular NASA's disparate functions as 'innovator and technology developer' vs. 'designer/developer/smart buyer' of new systems, and 'system operator' vs. 'service customer'.

I have my own opinions on these matters, as do many others in the space policy community, and am pleased to share them if asked. Some of those opinions I hold in common with some members of the Commission; in other cases 'not so much'. But the larger point is that these matters of national policy remain unsettled. I am truly gratified to see such substantive matters being raised by the Commission. They deserve correspondingly substantive debate, followed by decisive action.

So, at the strategic level, I believe that the Commission has done an excellent job of raising issues that matter and providing clear indications as to what the worthy and proper course for the nation's future in space should be.

At the same time, however, the Commission also addresses numerous tactical issues concerning how to go about achieving the goals they support, and offers views as to the merits of various implementation approaches considered during their deliberations. I think it is fair to say that I am less enamored of their treatment of these tactical issues than I am of their strategic assessments.

I believe that this is an important distinction to make, and that both strategy and tactics are important. Non-specialists will, and should, place great weight on the findings of this Commission. Where key tactical assessments and findings are at variance with those of knowledgeable and experienced practitioners, it can result in a level of public discord such that it becomes difficult for policy makers to know how to proceed. Thus, it will be important to consider carefully many specific points which were addressed by the Commission before decisions are made by the president and finally codified into law— or not — by the Congress.

The Commission notes, correctly, that NASA's Constellation program followed a design-to-cost strategy according to the budget profile of FY2005. NASA's budget as stipulated in 2005 was essentially constant in real dollars, with only a slight increase above inflation. Since then, it has since suffered some $30 billion of reductions to the amount allocated to human lunar return, including $12 billion in just the last five fiscal years.

The Commission notes that "Given the funding originally expected, the Constellation Program was a reasonable architecture for human exploration." In an earlier public statement, Commissioner Sally Ride noted that, "the program comes pretty close to performing as NASA advertised it would. ... NASA's planning and development phase of Constellation was actually pretty good." A veteran of the investigations of both the *Challenger* and *Columbia* accidents, Sally has seen her share of troubled programs, and so this comment was one I found telling.

Thus, one wonders why the Commission failed to recommend as its favored option the simplest one possible -- providing the funding necessary to do the job. Of all the options considered, this is the most straightforward. Yet it was not recommended. Other options are possible, of course, and the Commission would have been remiss not to explore them as well. But not to include this one is, in my view, simply wrong.

I say this because the civil space policy of the United States; e.g., "what NASA does", has been a matter of *law* since the passage in December 2005 of the NASA Authorization Act. This came about only after a full 23 months of fulsome, healthy, and productive debate on the merits of President George W. Bush's announcement of the "Vision for Exploration" in January 2004. The "Vision" itself was a response to another presidential commission, the Columbia Accident Investigation Board, which noted as a root cause of the Columbia accident the lack of a longterm strategic vision for NASA— a finding which I supported then and support now. In my view, the Congress extended and improved upon the original "Vision" in passing that Act, and did so again in 2008. On both occasions Congressional support for NASA's direction was heavily bipartisan.

Thus, when President Obama took office in January 2009, he inherited a civil space policy which had, in its essentials, survived six years of vigorous scrutiny, a space agency which had transformed itself to execute that policy, and could do so in a reasonable (if not very aggressive) timeframe on a constant-dollar budget as stipulated in 2005. The

Commission itself speaks of the need for stability in direction and funding, if NASA is to make reasonable progress and to be accountable for so doing. In my view, then, the most important question that Congress could ask of the new Administration and its Commission is this: exactly why does the policy which we have established in law — twice! — need to be changed?

We cannot discuss the civil space budget, budget stability, or future plans for human spaceflight without also addressing the future of the ISS. Certainly, the Commission fully recognized this point in their deliberations and in their Summary Report. However, the report devotes considerable attention to the issue of potentially decommissioning the ISS in 2016, trading the funds required for its extension against those required for the expansion of human spaceflight beyond LEO.

I must be clear. In my opinion, any discussion of decommissioning and deorbiting the ISS is irrelevant to the consideration of serious programmatic options. While it is certainly true that the Bush Administration did not provide funding for ISS past 2015, it was always quite clear that the decision to cancel or fund the ISS in 2016 and beyond was not within the purview of that administration to make. In the face of strong International Partner commitment to ISS and two decades of steadfast Congressional commitment to the ISS, it has never been and is not now realistic to consider decommissioning it in 2015, or indeed on *any* particular date which can be known today. The United States will not take unilateral action to cancel an international program which is the centerpiece of human spaceflight in every one of its fifteen participating nations, just because a particular date arrives on the calendar.

It has long been known that some $3+ billion per year will be required to sustain ISS operations past 2015. Failure to plan for this is, and has been, a glaring omission in the nation's budgetary policy. Thus, sustained funding of the ISS as long as it continues to return value— certainly to 2020 and quite likely beyond — should have been established by the Commission as a nonnegotiable point of departure for all other discussions.

The United States is now the majority owner of a 450 ton laboratory in space, a facility without compare. The fact that it took too long to build and that we spent more money on it than we should have is irrelevant to future decisions. We have it. We should use it to the maximum possible extent, for as long as we can make it last. But we must also go beyond ISS. The existence of future exploration programs cannot be traded against sustenance of the ISS on an "either-or" basis, as if that were a realistic option. If the nation is to have a viable human spaceflight program, *the requirement to sustain ISS while also developing new systems to go beyond low Earth orbit is the minimally necessary standard.* If the nation can no longer meet that standard, then it should be so stated, in which case any further discussion of U.S. human exploration beyond LEO is moot for the next two decades.

The Commission correctly addresses, front and center, concerns about the looming "gap" in independent U.S. access to LEO and to the ISS after the Space Shuttle is retired. To deal with this problem, the preference for "commercial" options for cargo and crew delivery to low Earth orbit appears throughout the Summary, together with the statement that "it is an appropriate time to consider turning this transport service over to the commercial sector." It must be asked: what commercial sector?

At present, the only clearly available "commercial" option to lift Orion as designed is the European Ariane 5, designed from the outset to be human rated. Even so, Arianespace has estimated that several years would be required to prepare the Ariane 5 and its processing infrastructure to meet the demands of human spaceflight. I believe this to be correct.

Launching a redesigned Orion crew vehicle on Ariane 5 is certainly a valid choice in the context of an international program. However, as an alternative to an independent U.S. government capability for human transport to LEO, it is a valid choice if, and *only* if, the U.S. is willing to give up independent access to low Earth orbit, a decision imbued with enormous future consequences. Are we really ready to take that step?

With an appropriately enlightened U.S. government policy there may one day be a domestic commercial space transportation sector. Such a policy could, as the Commission correctly notes, follow along the path laid out by government sponsorship of commercial air transportation in the last century (for cargo, by the way, not passenger traffic). No one in the space community wants that capability to exist more than I. But it does not presently exist, and will not exist in the near future; i.e., substantially prior to the expected availability of Ares-1 and Orion, if properly funded.

The key point is this: the existence of a guaranteed U.S. government option for cargo and crew delivery to ISS is what allows government to take prudent risks to help bring about the development of a viable commercial space sector.

The Commission acknowledges the "risk" associated with its recommendation, but is not clear about the nature of that risk. If no government option to deliver cargo and crew to LEO is developed following the retirement of the Space Shuttle, the U.S. risks the failure to sustain and utilize a unique facility with a sunk cost of $55 billion on the U.S. side, and nearly $20 billion of international partner investment. The Russian Soyuz and Progress systems, even if we are willing to be dependent upon Russia and are willing to pay whatever is required for their use, simply do not provide sufficient capability to utilize ISS as was intended. Further, they represent a single point failure in regard to such utilization. In my view, to hold the support and utilization of the ISS hostage to the emergence of a commercial space sector is not "risky", it is irresponsible.

The Commission claims that safety "is not discussed in extensive detail because any concepts falling short in human safety have simply been eliminated from consideration." Similarly, the Commission was "unconvinced that enough is known about any of the potential high-reliability launcher-plus-capsule systems to distinguish their levels of safety in a meaningful way." For those of us in the human spaceflight community, this is a "hot button". The Commission has dismissed out of hand the extensive work that has been done to assure that Constellation systems offer the safest approach in comparison to all other presently known systems. This is simply unacceptable. Work of high quality in the assessment of safety and reliability has been done, and independently validated discriminators between and among various system options do exist, whether the Commissioners believe so or not. Further, the Summary Report is confusing as regards the distinction between "reliability" and "safety", where it is discussed at all. The former is the only criterion of interest for unmanned systems; for manned systems, there is an important difference due to the existence of an abort system and the conditions under which that abort system can and must operate. Nowhere is this crucial distinction discussed.

The Commission recommends consideration of a lunar mission architecture featuring a dual- launch of the Ares-5 Lite vehicle, instead of the Ares-1/Ares-5 Constellation baseline. The rationale for this recommendation is difficult to understand, because economic considerations favor Ares-5 over Ares-5 Lite. Ares-5 costs 12% more to develop than Ares-5 Lite, but carries 14% more payload to LEO and 20% more payload to the moon (50 mt vs. 60

mt). Even more importantly, the operations cost for the dual-Ares-5 Lite lunar mission concept is several hundred million dollars higher than the baseline plan, for the same reference program of two human and two cargo missions to the moon each year.

The Commission agrees that a heavy-lift launcher is needed for human space exploration beyond LEO. Because of the economies of scale inherent to the design of launch vehicles, the cost-per-pound of payload to orbit nearly always improves with increasing launch vehicle size. Thus, a heavy-lift vehicle should be designed to be as large as possible within the constraints of the facilities and infrastructure available to build and transport it. This provides the greatest marginal capability at the lowest marginal cost.

The use of "fuel depots" as recommended in the Summary Report is equally difficult to understand. The Ares-5 offers the lowest cost-per-pound for payload to orbit of any presently known launch vehicle design. An architectural approach based upon the use of numerous smaller vehicles to stock a fuel depot is inevitably more expensive than putting the necessary payload up in larger pieces. Further, a fuel depot requires a presently non-existent technology — the ability to maintain cryogenic fuels in the necessary thermodynamic state for very long periods in space. This technology is a holy grail of deep-space exploration, because it is necessary for both chemical- and nuclear-powered upper stages. To embrace an architecture based upon a non-existent technology at the very beginning of beyond-LEO operations is unwise.

Finally, there are a number of concerns as to the methodology by which the Commission reached some of its conclusions.

When trying to assess the relative merits of multiple options for an engineering design — in this case the design of spaceflight architectures—the core requirement to allow meaningful comparisons is to fix the goals and constraints so that these "boundary conditions" are common for all. In the Commission's report, various options are presented which are not linked by common goals and constraints. Instead, differing options with different constraints are presented to reach disparate goals, rendering it impossible to develop meaningful cost/schedule/performance/risk comparisons across them. These options possess vastly differing levels of maturity, yet are offered as if all were on an equal footing in regard to their level of technical, cost, schedule, and risk assessment.

Significantly, *no* trade study was performed to assess how well each of the options considered by the Commission performed in meeting the goals and constraints of the existing U.S. civil space policy, as it is governed by the NASA Authorization Acts of 2005 and 2008.

The Commission cites "independent" cost estimates for Constellation systems. There is no acknowledgement that these are low-fidelity estimates developed over a matter of a few weeks, yet are offered as corrections to NASA's cost estimates, which have years of rigorou effort behind them.

As one example, it is common in cost analysis to apply a large historical cost growth factor to preliminary estimates for new designs. The size of the factor depends on the nature of the work being done and the maturity of the original estimate. Work done by Aerospace Corporation to model cost growth in certain classes of robotic space systems suggests that a growth factor of about 50% might be appropriate for the design and development of a new system. Hence, that factor was applied to the assessment of "clean sheet" options offered to the Commission. However, the same factor was also applied to NASA's Constellation element designs. This is, effectively, "double counting". Historical growth factors were incorporated into Constellation costs from the very first, and are reflected in delivery schedule projections for the various system elements, Ares-1, Orion, etc. To apply a new "growth factor" on top ofthose in the original models is misleading.

The Commission does not acknowledge NASA's commitment to probabilistic budget estimation techniques for Constellation, at a 65% cost-confidence level — higher than has ever been the case in the past. This is a fundamental break from past practice at the agency, a key to providing more realistic information on program status to agency managers and external stakeholders.

If the Commission believes that NASA is not using state-of-the-art methodologies to estimate costs, or is misrepresenting the data it has amassed, it should document its specific concerns. Otherwise, the provenance of NASA's cost estimates should be accepted, as no evidence has been supplied to justify overturning them.

the severity of such problems. Constellation's "technical problems" are on display because actual work is being accomplished. Other options have no problems because no work is being done. There are never any technical problems on viewgraphs.

To this point, in *The Rickover Effect: How One Man Made a Difference*, Theodore Rockwell recalls a priceless observation by Adm. Hyman Rickover. When confronted with a situation in which a variety of alternative concepts were being advocated to— and around — Rickover in place of the pressurized-water reactor design he favored for the nuclear navy, Rickover noted that there were two kinds of reactors, "paper reactors"; i.e., new reactor concepts, and "real reactors". A paper reactor has the following characteristics:

- It is simple.
- It is small.
- It is cheap.
- It is lightweight.
- It can be built very quickly.
- Very little development is required; it can use off-the-shelf components.
- It is in the study phase; it is not being built now.

In contrast, a real reactor has the following characteristics:

- It is complicated.
- It is large.
- It is heavy.
- It is being built now.
- It is behind schedule.
- It requires an immense amount of development on apparently trivial items.
- It takes a long time to build because of its engineering development problems.

Does any of this sound familiar?

Finally, the Commission did not do that which would have been most valuable— rendering a clear-eyed, independent assessment of the progress and status of Constellation with respect to its ability to meet the goals which have been established in two successive NASA Authorization Acts, followed by an assessment of what would be required to get and keep that program on track. Instead, the Commission sought to formulate new options for new programs, treating these options as if their level of maturity was comparable to that of the baseline upon which NASA has been working now for more than four years. This ignores the established body of law which has guided NASA's work for the last four years and which, until and unless that body of law is changed, must serve as the common standard for any proposed alternative to Constellation as the "program of record" for the nation's existing human spaceflight program.

With the above having been said, where do we go from here? In the end that is the only important question. Let me be as clear as possible on a further point. When I noted above that the best option is to restore funding, I do not want to mislead this Committee. It is not possible to recover fully, in terms of schedule, personnel morale, and programmatic decisions, from the damage which has been done to NASA and to Constellation by reductions in funding, particularly in the last couple of years, when the program has moved into full-bore execution. Past decisions and actions are a form of sunk cost. So I do not propose to render the program somehow magically "whole" by restoring past funding cuts. That cannot be done. But NASA does know — or can shortly assess — what is necessary to get Constellation back on track with regard to the best achievable schedule, *from where we are today*, for regaining access to LEO, returning to the Moon, exploring some of the near-Earth asteroids, and eventually voyaging to Mars.

The details will, as I say, best come from NASA. However, I can suggest what I think might be the most viable alternative if we remain committed both to continuing ISS operations and to human exploration beyond LEO, yet cannot return all of the money to the NASA budget that has been removed in the last few years.

In such a case, at least in my opinion, it would be logical to delay lunar lander development in order to make progress on the other elements. I don't think it is a very good idea to try to make it "smaller" or somehow less capable in some other way. Current planning is for a crew of four on the moon. Carrying two pairs of two EVA crewmembers is very logical, for all the reasons that apply to Shuttle today. It also has the advantage of providing ample opportunities for crew from international partners.

If that rationale is accepted, then I think it makes more sense to delay the lander development than to compromise the design of a machine that will be in use for a very long time. So, the Altair lunar lander would be built when the money to do so becomes available. Ares-1 and Orion should be completed as quickly as possible to support ISS, and then Ares-5 should be built. They should not all be done in parallel; that causes them to stretch out and costs more in the long run. It makes more sense to start some elements later. In the meantime, once Ares 5 becomes available but prior to human lunar return, Orion could be used for some of the "Flexible Path" options cited by the Commission. Such options were, in fact, considered from the first during ESAS. The use of Constellation hardware for destinations that were not included in the Vision for Space Exploration (as initially stated) was a core part of our thinking during ESAS. I considered that to be a strong point of the chosen architecture — it was flexible about destinations. An Orion spacecraft that can take care of itself for six months around the moon can go a lot of other places.

I think that some variant of the approach outlined above makes the most sense going forward. It would position us as well for the future as we can be, given where we are today, unless a substantial sum of money can be allocated to the original plan for lunar return by 2020.

The Summary Report suggests inviting international partners into the critical path of program development. This is a valid alternative if we are willing to depart significantly from prior policy. Europe, Japan, or Russia could build a lunar lander just as well as could the United States. Politically and culturally, this would be a big step. I sat in front of this Committee, with a different Chairman, when former Administrator Dan Goldin was advised in very direct terms to "keep Russia off the critical path on the ISS". But, if we wanted to be more "inclusive", we could decide that the United States will develop the heavy-lift launcher and deep-space crew vehicle, but a return to the moon will depend upon international partner contributions. I personally do not favor such an approach, but it is a technically feasible option.

I would like to close with a quote from the Commission's Summary Report: "Finally, significant space achievements require continuity of support over many years. One way to ensure that no successes are achieved is to continually pull up the flowers to see if the roots are healthy. (This Committee might be accused of being part of this pattern!)"

I couldn't agree more. As I see it, the Commission didn't find anything wrong with the current program, didn't find anything safer, more reliable, cheaper or faster. The roots are healthy. So, why throw away four years and $8 billion pulling up the flowers? Let's apply some plant nutrient and watch them grow.

This, to me, is our best option for re-affirming a stable civil space policy. Thank you.

In: Future of U.S. Human Spaceflight: Background and Issues ISBN: 978-1-61668-774-8
Ed: Derek A. Warren and Bridget D. Conway

Chapter 7

STATEMENT OF JOHN D. ROCKEFELLER, IV HEARING: OPTIONS FROM THE REVIEW OF U.S. HUMAN SPACE FLIGHT PLANS COMMITTEE*

I want to thank Mr. Augustine for testifying before the Subcommittee today and I want to express my gratitude to him and the other 9 members of the Human Space Flight Plans Committee for their service.

The task of examining NASA's human space flight plans in a mere 90 days is not an easy one. As the committee states in its Summary Report, it is in fact rocket science. Thank you for your commitment to the process and its smooth progress.

I am concerned, however, that the committee seems to have largely accepted the value of human space exploration as an underlying assumption without ever fully discussing it.

In this difficult budget environment where the federal deficit sets new records with each passing month, our country does not necessarily have the resources to do everything that we want to do. The economy has not yet recovered, we are fighting two major wars, and we can't even manage to find a solution to provide healthcare for all Americans.

The review committee recommends increasing NASA's budget by an additional $3 billion by FY2014, then increasing it at an inflationary rate of 2.4%. We must ask, quite simply, what are the specific benefits to the nation of making NASA a $22 billion agency?

Let me be clear, I am not advocating that we abandon space exploration. The case, however, has not yet been made. And I think now would be an appropriate time to ask those tough questions and carefully evaluate the responses.

This is no longer the era of Apollo and the Cold War where the payoffs for advancing the space and Moon agenda are entirely clear.

The president, I am sure, will continue to struggle with these same questions as he reviews the committee's findings and options. I am committed to making sure that together we can get the answers right.

Again, I want to thank Mr. Augustine for his service and testimony today. I look forward to the committee's final report.

---

* This is an edited, reformatted and augmented edition of statement/testimony given before the House of Representatives on Septmber 16, 2009.

INDEX

C

D

G

H

I

J

K

L

M

N

O

P

Q

R

T

U

V

W

Y